Learn

Eureka Math®

Grade 3
Module 7

Published by Great Minds®.

Printed in the U.S.A.
This book may be purchased from the publisher at eureka-math.org.
BAB 10 9 8 7 6 5 4 3

ISBN 978-1-64054-063-7

G3-M7-L-05.2018

Learn • Practice • Succeed

Eureka Math® student materials for *A Story of Units®* (K–5) are available in the *Learn, Practice, Succeed* trio. This series supports differentiation and remediation while keeping student materials organized and accessible. Educators will find that the *Learn, Practice,* and *Succeed* series also offers coherent—and therefore, more effective—resources for Response to Intervention (RTI), extra practice, and summer learning.

Learn

Eureka Math Learn serves as a student's in-class companion where they show their thinking, share what they know, and watch their knowledge build every day. *Learn* assembles the daily classwork—Application Problems, Exit Tickets, Problem Sets, templates—in an easily stored and navigated volume.

Practice

Each *Eureka Math* lesson begins with a series of energetic, joyous fluency activities, including those found in *Eureka Math Practice.* Students who are fluent in their math facts can master more material more deeply. With *Practice,* students build competence in newly acquired skills and reinforce previous learning in preparation for the next lesson.

Together, *Learn* and *Practice* provide all the print materials students will use for their core math instruction.

Succeed

Eureka Math Succeed enables students to work individually toward mastery. These additional problem sets align lesson by lesson with classroom instruction, making them ideal for use as homework or extra practice. Each problem set is accompanied by a Homework Helper, a set of worked examples that illustrate how to solve similar problems.

Teachers and tutors can use *Succeed* books from prior grade levels as curriculum-consistent tools for filling gaps in foundational knowledge. Students will thrive and progress more quickly as familiar models facilitate connections to their current grade-level content.

Students, families, and educators:

Thank you for being part of the *Eureka Math*® community, where we celebrate the joy, wonder, and thrill of mathematics.

In the *Eureka Math* classroom, new learning is activated through rich experiences and dialogue. The *Learn* book puts in each student's hands the prompts and problem sequences they need to express and consolidate their learning in class.

What is in the Learn book?

Application Problems: Problem solving in a real-world context is a daily part of *Eureka Math.* Students build confidence and perseverance as they apply their knowledge in new and varied situations. The curriculum encourages students to use the RDW process—Read the problem, Draw to make sense of the problem, and Write an equation and a solution. Teachers facilitate as students share their work and explain their solution strategies to one another.

Problem Sets: A carefully sequenced Problem Set provides an in-class opportunity for independent work, with multiple entry points for differentiation. Teachers can use the Preparation and Customization process to select "Must Do" problems for each student. Some students will complete more problems than others; what is important is that all students have a 10-minute period to immediately exercise what they've learned, with light support from their teacher.

Students bring the Problem Set with them to the culminating point of each lesson: the Student Debrief. Here, students reflect with their peers and their teacher, articulating and consolidating what they wondered, noticed, and learned that day.

Exit Tickets: Students show their teacher what they know through their work on the daily Exit Ticket. This check for understanding provides the teacher with valuable real-time evidence of the efficacy of that day's instruction, giving critical insight into where to focus next.

Templates: From time to time, the Application Problem, Problem Set, or other classroom activity requires that students have their own copy of a picture, reusable model, or data set. Each of these templates is provided with the first lesson that requires it.

Where can I learn more about Eureka Math *resources?*

The Great Minds® team is committed to supporting students, families, and educators with an ever-growing library of resources, available at eureka-math.org. The website also offers inspiring stories of success in the *Eureka Math* community. Share your insights and accomplishments with fellow users by becoming a *Eureka Math* Champion.

Best wishes for a year filled with aha moments!

Jill Diniz

Jill Diniz
Director of Mathematics
Great Minds

The Read–Draw–Write Process

The *Eureka Math* curriculum supports students as they problem-solve by using a simple, repeatable process introduced by the teacher. The Read–Draw–Write (RDW) process calls for students to

1. Read the problem.
2. Draw and label.
3. Write an equation.
4. Write a word sentence (statement).

Educators are encouraged to scaffold the process by interjecting questions such as

- What do you see?
- Can you draw something?
- What conclusions can you make from your drawing?

The more students participate in reasoning through problems with this systematic, open approach, the more they internalize the thought process and apply it instinctively for years to come.

Contents

Module 7: Geometry and Measurement Word Problems

Topic E: Problem Solving with Perimeter and Area

Topic F: Year in Review

Name ______________________________ Date ______________

Lena's family visits Little Tree Apple Orchard. Use the RDW process to solve the problems about Lena's visit to the orchard. Use a letter to represent the unknown in each problem.

1. The sign below shows information about hayrides at the orchard.

Hayrides

Adult ticket $7

Child ticket $4

Leaves every 15 minutes starting at 11:00.

a. Lena's family buys 2 adult tickets and 2 child tickets for the hayride. How much does it cost Lena's family to go on the hayride?

b. Lena's mom pays for the tickets with $5 bills. She receives $3 in change. How many $5 bills does Lena's mom use to pay for the hayride?

c. Lena's family wants to go on the fourth hayride of the day. It's 11:38 now. How many minutes do they have to wait for the fourth hayride?

2. Lena picked 17 apples, and her brother picked 19. Lena's mom has a pie recipe that requires 9 apples. How many pies can Mom make with the apples that Lena and her brother picked?

3. Lena's dad gives the cashier $30 to pay for 6 liters of apple cider. The cashier gives him $6 in change. How much does each liter of apple cider cost?

4. The apple orchard has 152 apple trees. There are 88 trees with red apples. The rest of the trees have green apples. How many more trees have red apples than green apples?

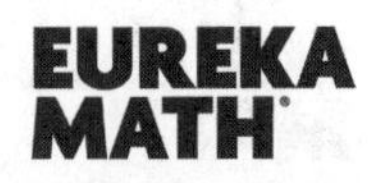

Name ______________________________ Date ______________

Use the RDW process to solve the problem below. Use a letter to represent the unknown.

Sandra keeps her sticker collection in 7 albums. Each album has 40 stickers in it. She starts a new album that has 9 stickers in it. How many total stickers does she have in her collection?

Name ______________________________________ Date ____________________

Use the RDW process to solve. Use a letter to represent the unknown in each problem.

1. Leanne needs 120 tiles for an art project. She has 56 tiles. If tiles are sold in boxes of 8, how many more boxes of tiles does Leanne need to buy?

2. Gwen pours 236 milliliters of water into Ravi's beaker. Henry pours 189 milliliters of water into Ravi's beaker. Ravi's beaker now contains 800 milliliters of water. How much water was in Ravi's beaker to begin with?

3. Maude hung 3 pictures on her wall. Each picture measures 8 inches by 10 inches. What is the total area of the wall covered by the pictures?

4. Kami scored a total of 21 points during her basketball game. She made 6 two-point shots, and the rest were three-point shots. How many three-point shots did Kami make?

5. An orange weighs 198 grams. A kiwi weighs 85 grams less than the orange. What is the total weight of the fruit?

6. The total amount of rain that fell in New York City in two years was 282 centimeters. In the first year, 185 centimeters of rain fell. How many more centimeters of rain fell in the first year than in the second year?

Name ______________________________________ Date ________________

Use the RDW process to solve the problem below. Use a letter to represent the unknown.

Jaden's bottle contains 750 milliliters of water. He drinks 520 milliliters at practice and then another 190 milliliters on his way home. How many milliliters of water are left in Jaden's bottle when he gets home?

Name ______________________________ Date ______________

Use the RDW process to solve the problems below. Use a letter to represent the unknown in each problem. When you are finished, share your solutions with a partner. Discuss and compare your strategies with your partner's strategies.

1. Monica measures 91 milliliters of water into 9 tiny beakers. She measures an equal amount of water into the first 8 beakers. She pours the remaining water into the ninth beaker. It measures 19 milliliters. How many milliliters of water are in each of the first 8 beakers?

2. Matthew and his dad put up 8 six-foot lengths of fence on Monday and 9 six-foot lengths on Tuesday. What is the total length of the fence?

3. The total weight of Laura's new pencils is 112 grams. One pencil rolls off the scale. Now the scale reads 105 grams. What is the total weight of 7 new pencils?

4. Mrs. Ford's math class starts at 8:15. They do 3 fluency activities that each last 4 minutes. Just when they finish all of the fluency activities, the fire alarm goes off. When they return to the room after the drill, it is 8:46. How many minutes did the fire drill last?

5. On Saturday, the baker bought a total of 150 pounds of flour in five-pound bags. By Tuesday, he had 115 pounds of flour left. How many five-pound bags of flour did the baker use?

6. Fred cut an 84-centimeter rope into 2 parts and gave his sister 1 part. Fred's part is 56 centimeters long. His sister cut her rope into 4 equal pieces. How long is 1 of his sister's pieces of rope?

Name ______________________________ Date ______________

Use the RDW process to solve the problem below. Use a letter to represent the unknown.

Twenty packs of fruit snacks come in a box. Each pack weighs 6 ounces. Students eat some. There are 48 ounces of fruit snacks left in the box. How many ounces of fruit snacks did the students eat?

Student A

Total pencils

9	9	9	9	9	9

$6 \times 9 = 54$

Pencils she gave away

24×2

$(6 \times 4) \times 2$

$6 \times (4 \times 2)$

$6 \times 8 = 48$

$54 - 48 = 6$

Mrs. Mashburn has 6 pencils left.

Student B

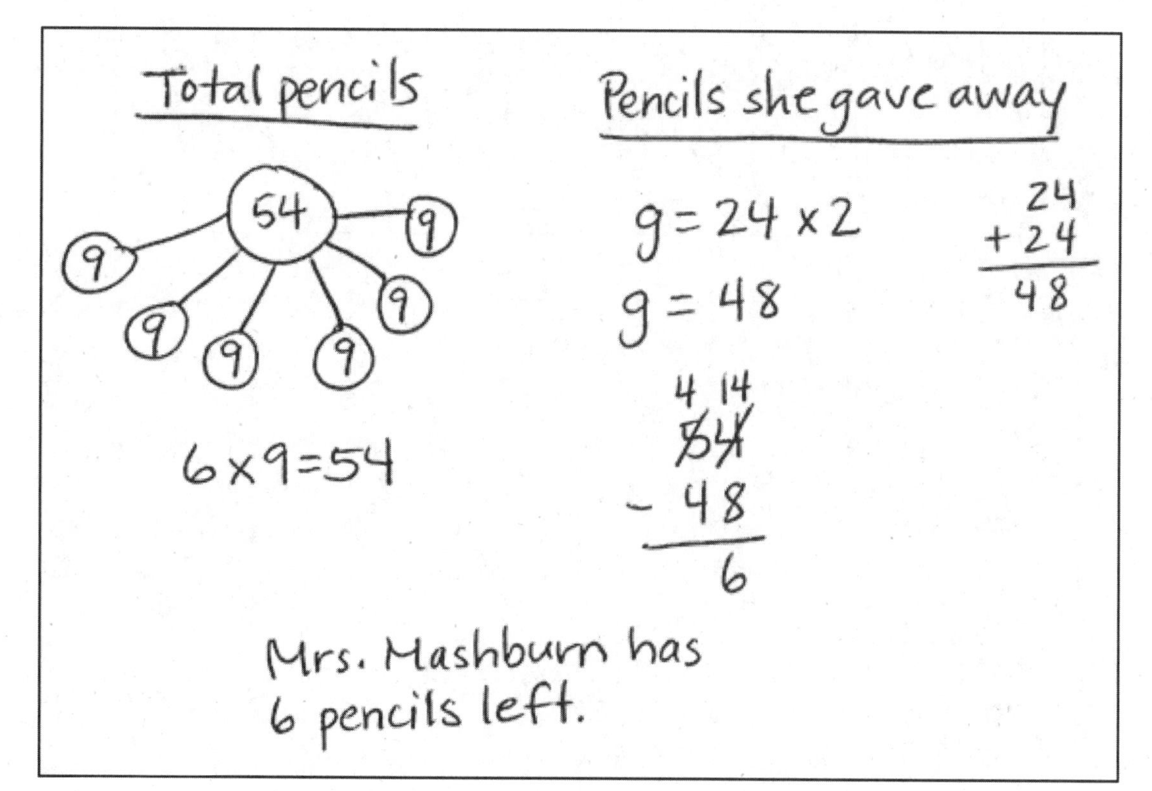

---

student work samples

Student C

$$\begin{array}{r} \overset{4}{\cancel{5}}\overset{14}{\cancel{4}} \\ -48 \\ \hline 06 \end{array}$$

Mrs. Mashburn has 6 pencils left.

student work samples

The third graders raised $437 in a fundraiser. The fourth graders raised $68 less than the third graders. How much money did the two grade levels raise altogether?

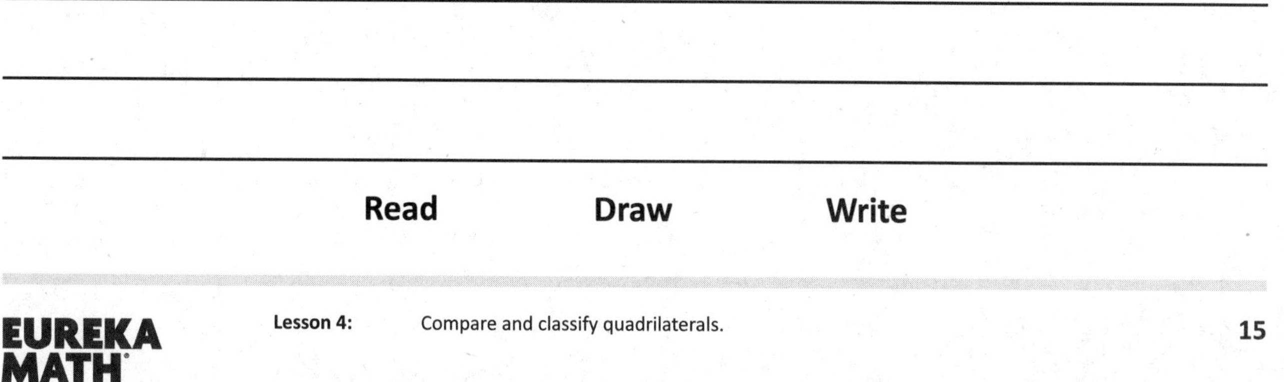

Name ____________________ Date ____________

1. Cut out all the polygons (A–L) in the Template. Then, use the polygons to complete the following chart.

Attribute	Write the letters of the polygons in this group.	Sketch 1 polygon from the group.
Example: **3 Sides**	Polygons: Y, Z	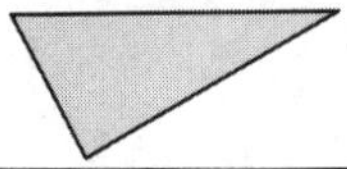
4 Sides	Polygons:	
At Least 1 Set of Parallel Sides	Polygons:	
2 Sets of Parallel Sides	Polygons:	
4 Right Angles	Polygons:	
4 Right Angles and 4 Equal Sides	Polygons:	

2. Write the letters of the polygons that are quadrilaterals. Explain how you know these polygons are quadrilaterals.

3. Sketch a polygon below from the group that has 2 sets of parallel sides. Trace 1 pair of parallel sides red. Trace the other pair of parallel sides blue. What makes parallel sides different from sides that are not parallel?

4. Draw a diagonal line from one corner to the opposite corner of each polygon you drew in the chart using a straightedge. What new polygon(s) did you make by drawing the diagonal lines?

Name ________________________________ Date ________________

List as many attributes as you can to describe each polygon below.

1.

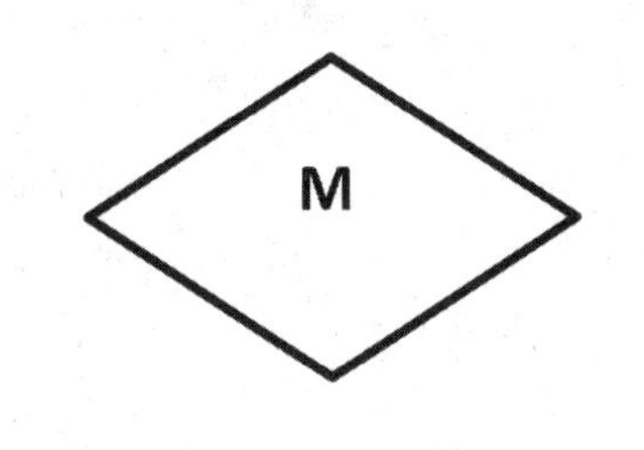

2.

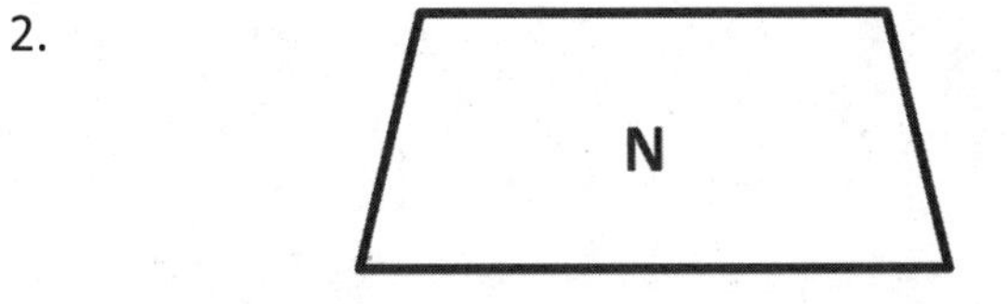

Name ______________________________ Date ______________

1. Cut out all the polygons (M–X) in the Template. Then, use the polygons to complete the following chart.

Attribute	List polygons' letters for each group.	Sketch 1 polygon from the group.
Example: **3 Sides**	Polygons: Y, Z	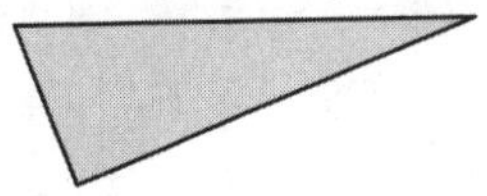
All Sides Are Equal	Polygons:	
All Sides Are Not Equal	Polygons:	
At Least 1 Right Angle	Polygons:	
At Least 1 Set of Parallel Sides	Polygons:	

2. Compare Polygon M and Polygon X. What is the same? What is different?

3. Jenny says, "Polygon N, Polygon R, and Polygon S are all regular quadrilaterals!" Is she correct? Why or why not?

4. "I have six equal sides and six equal angles. I have three sets of parallel lines. I have no right angles."

   a. Write the letter and the name of the polygon described above.

   b. Estimate to draw the same type of polygon as in part (a), but with no equal sides.

Name ____ Date ____________________

Jonah draws the polygon below. Use your ruler and right angle tool to measure his polygon. Then, answer the questions below.

1. Is Jonah's polygon a regular polygon? Explain how you know.

2. How many right angles does his polygon have? Circle the right angles on his polygon.

3. How many sets of parallel lines does his polygon have?

4. What is the name of Jonah's polygon?

Frankie says that all squares are rectangles, but not all rectangles are squares. Do you agree with this statement? Why or why not? Draw diagrams to support your statement.

Read Draw Write

Name ______________________________ Date ______________

Use a ruler and a right angle tool to help you draw the figures with the attributes given below.

1. Draw a triangle with 1 right angle.

2. Draw a quadrilateral with 4 right angles and sides that are all 2 inches long.

3. Draw a quadrilateral with at least 1 set of parallel sides. Trace the parallel sides green.

4. Draw a pentagon with at least 2 equal sides. Label the 2 equal side lengths of your shape.

5. Draw a hexagon with at least 2 equal sides. Label the 2 equal side lengths of your shape.

6. Sam says that he drew a polygon with 2 sides and 2 angles. Can Sam be correct? Use pictures to help you explain your answer.

Name ____ Date ____________________

Use a ruler and a right angle tool to help you draw a shape that matches the attributes of Jeanette's shape. Label your drawing to explain your thinking.

Jeanette says her shape has 4 right angles and 2 sets of parallel sides. It is not a regular quadrilateral.

Name ______________________________ Date ______________

1. Use tetrominoes to create at least two different rectangles. Then, color the grid below to show how you created your rectangles. You may use the same tetromino more than once.

2. Use tetrominoes to create at least two squares, each with an area of 36 square units. Then, color the grid below to show how you created your squares. You may use the same tetromino more than once.

   a. Write an equation to show the area of a square above as the sum of the areas of the tetrominoes you used to make the square.

   b. Write an equation to show the area of a square above as the product of its side lengths.

3. a. Use tetrominoes to create at least two different rectangles, each with an area of 12 square units. Then, color the grid below to show how you created the rectangles. You may use the same tetromino more than once.

b. Explain how you know the area of each rectangle is 12 square units.

4. Marco created a rectangle with tetrominoes and traced its outline in the space below. Use tetrominoes to re-create it. Estimate to draw lines inside the rectangle below to show how you re-created Marco's rectangle.

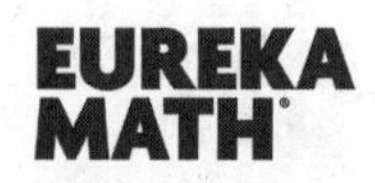

Name ______________________________ Date ______________

Use your tetrominoes to make a rectangle that has an area of 20 square units. Then, color the grid to show how you made your rectangle. You may use the same tetromino more than once.

Name ______________________________ Date ____________

1. Fold and cut the square on the diagonal. Draw and label your 2 new shapes below.

2. Fold and cut one of the triangles in half. Draw and label your 2 new shapes below.

3. Fold twice, and cut your large triangle. Draw and label your 2 new shapes below.

4. Fold and cut your trapezoid in half. Draw and label your 2 new shapes below.

5. Fold and cut one of your trapezoids. Draw and label your 2 new shapes below.

6. Fold and cut your second trapezoid. Draw and label your 2 new shapes below.

7. Reconstruct the original square using the seven shapes.

   a. Draw lines inside the square below to show how the shapes go together to form the square. The first one has been done for you.

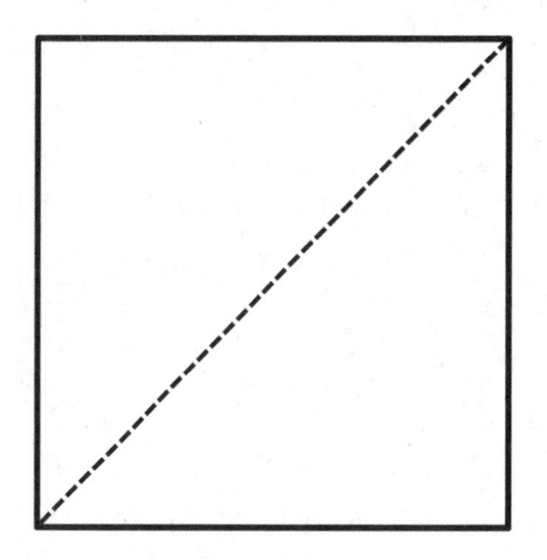

   b. Describe the process of forming the square. What was easy, and what was challenging?

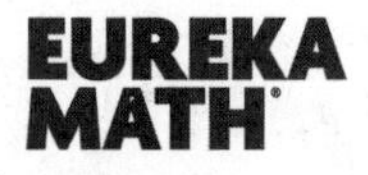

Name ______________________________ Date ______________

Choose three shapes from your tangram puzzle. Trace them below. Label the name of each shape, and describe *at least* one attribute that they have in common.

Name at least two attributes that a trapezoid, a square, and a parallelogram all have in common. Draw a diagram to support your ideas.

Read **Draw** **Write**

Name ______________________________ Date ______________

1. Use at least two tangram pieces to make and draw two of each of the following shapes. Draw lines to show where the tangram pieces meet.

   a. A rectangle that does not have all equal sides.

   b. A triangle.

   c. A parallelogram.

   d. A trapezoid.

2. Use your two smallest triangles to create a square, a parallelogram, and a triangle. Show how you created them below.

3. Create your own shape on a separate sheet of paper using all seven pieces. Describe its attributes below.

4. Trade your outline with a partner to see if you can re-create her shape using your tangram pieces. Reflect on your experience below. What was easy? What was challenging?

Name ______________________________________ Date ____________________

Nancy uses her tangram pieces to make a trapezoid without using the square piece. Below, sketch how she might have created her trapezoid.

Trista uses all seven of her tangram pieces to make a square as shown. One side of the large square is 4 inches long. What is the total area of the two large triangles? Explain your answer.

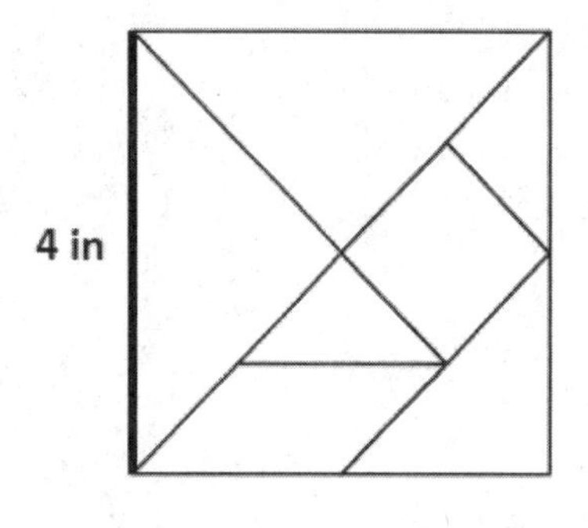

____

____

____

____

Read **Draw** **Write**

Name ____________________________________   Date ________________

1. Use a 2-inch square to answer the questions below.

   a. Trace the square in the space below with a red crayon.

   b. Trace the new shape you made with the square in the space below with a red crayon.

   c. Which shape has a greater perimeter? How do you know?

   d. Color the inside of the shapes in Problem 1 (a) and (b) with a blue crayon.

e. Which color represents the perimeters of the shapes? How do you know?

f. What does the other color represent? How do you know?

g. Which shape has a greater area? How do you know?

2. a. Outline the perimeter of the shapes below with a red crayon.

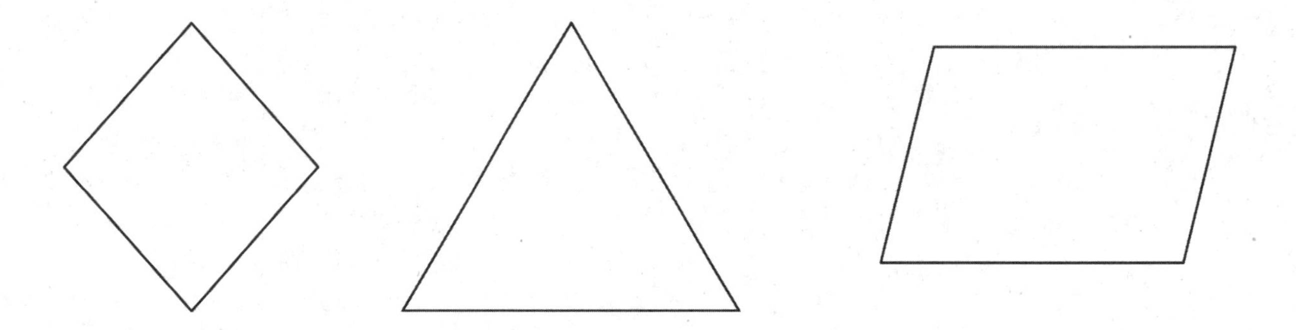

b. Explain how you know you outlined the perimeters of the shapes above.

3. Outline the perimeter of this piece of paper with a highlighter.

Name ______________________________ Date ______________

Jason paints the outside edges of a rectangle purple. Celeste paints the inside of the rectangle yellow.

1. Use your crayons to color the rectangle that Jason and Celeste painted.

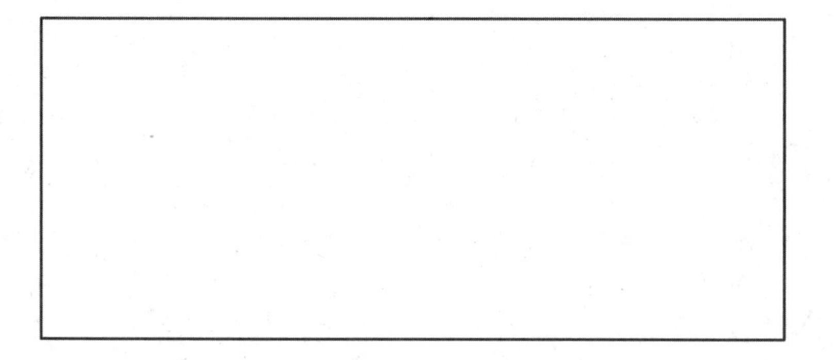

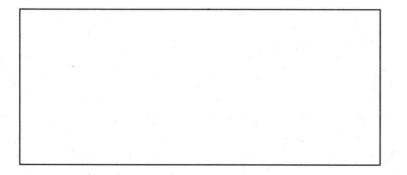

2. Which color represents the perimeter of the rectangle? How do you know?

Name ____________________________________ Date ________________

1. Follow the directions below using the shape you created yesterday.
   a. Tessellate your shape on a blank piece of paper.
   b. Color your tessellation to create a pattern.
   c. Outline the perimeter of your tessellation with a highlighter.
   d. Use a string to measure the perimeter of your tessellation.

2. Compare the perimeter of your tessellation to a partner's. Whose tessellation has a greater perimeter? How do you know?

3. How could you increase the perimeter of your tessellation?

4. How would overlapping your shape when you tessellated change the perimeter of your tessellation?

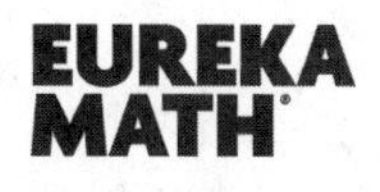

Name ____ Date ____________________

Estimate to draw at least four copies of the given regular hexagon to make a new shape, without gaps or overlaps. Outline the perimeter of your new shape with a highlighter. Shade in the area with a colored pencil.

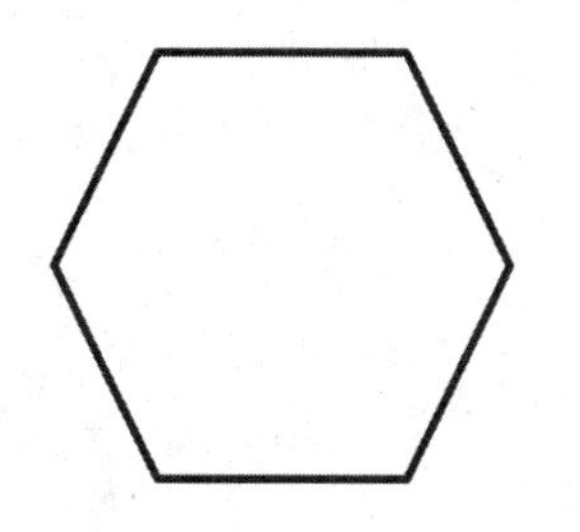

Angela measures the sides of a Square napkin with her ruler. Each side measures 6 inches. What is the perimeter of the napkin?

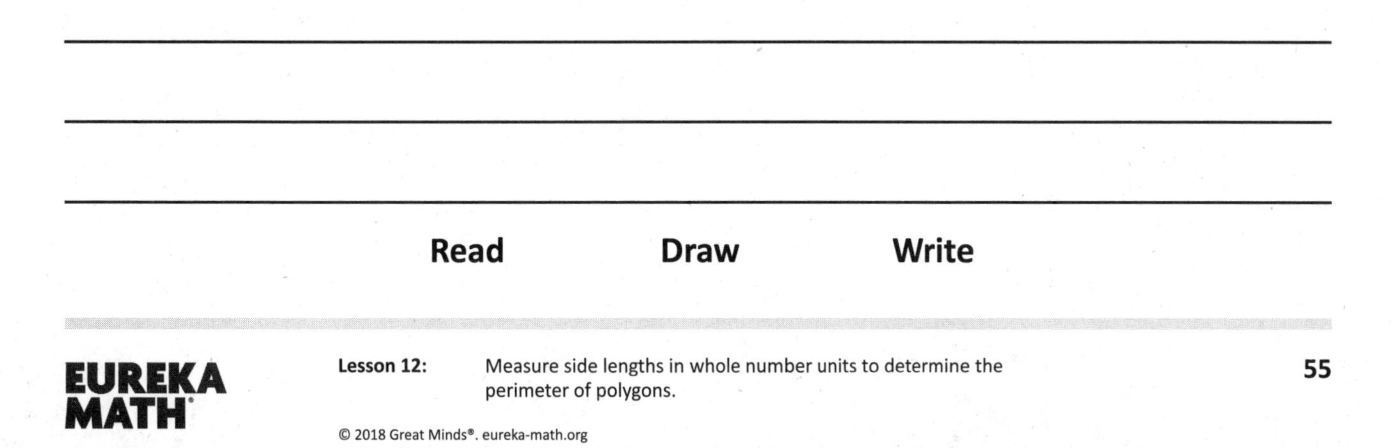

Name ______________________________ Date ______________

1. Measure and label the side lengths of the shapes below in centimeters. Then, find the perimeter of each shape.

a.

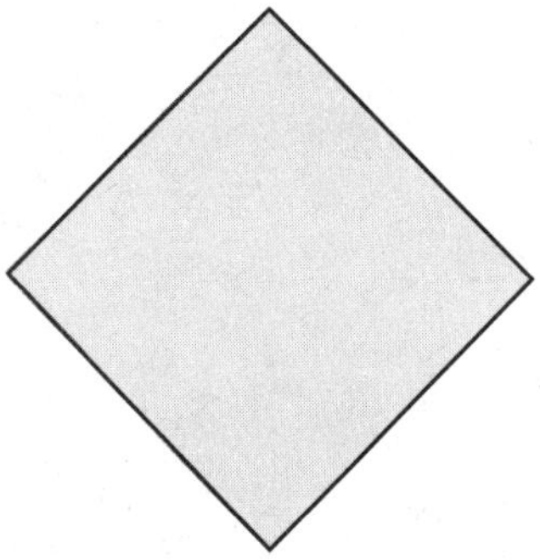

Perimeter = _____ cm + _____ cm + _____ cm + _____ cm

= _______ cm

b.

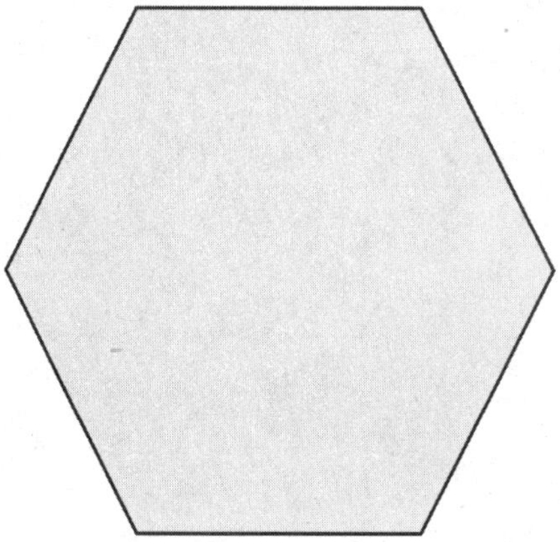

Perimeter = ____________________

= _______ cm

c.

Perimeter = ____________________

= _______ cm

d.

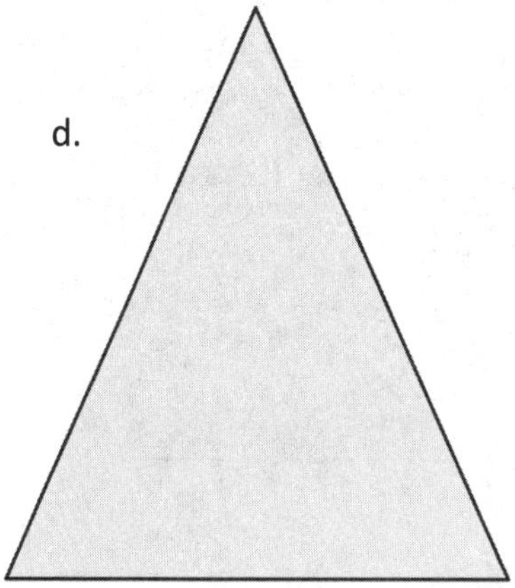

Perimeter = ____________________

= _______ cm

e.

Perimeter = ____________________

= _______ cm

2. Carson draws two triangles to create the new shape shown below. Use a ruler to find the side lengths of Carson's shape in centimeters. Then, find the perimeter.

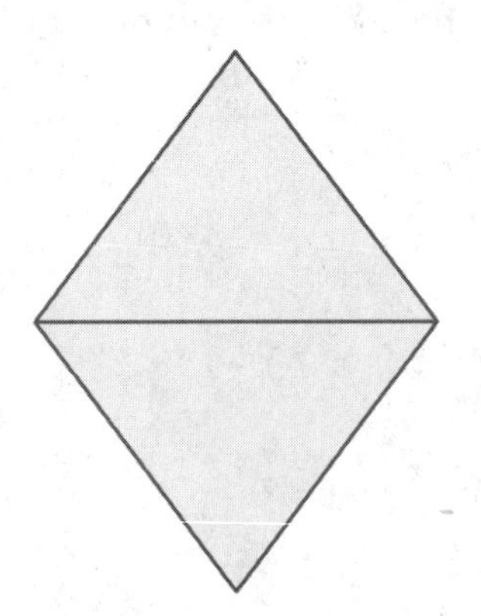

3. Hugh and Daisy draw the shapes shown below. Measure and label the side lengths in centimeters. Whose shape has a greater perimeter? How do you know?

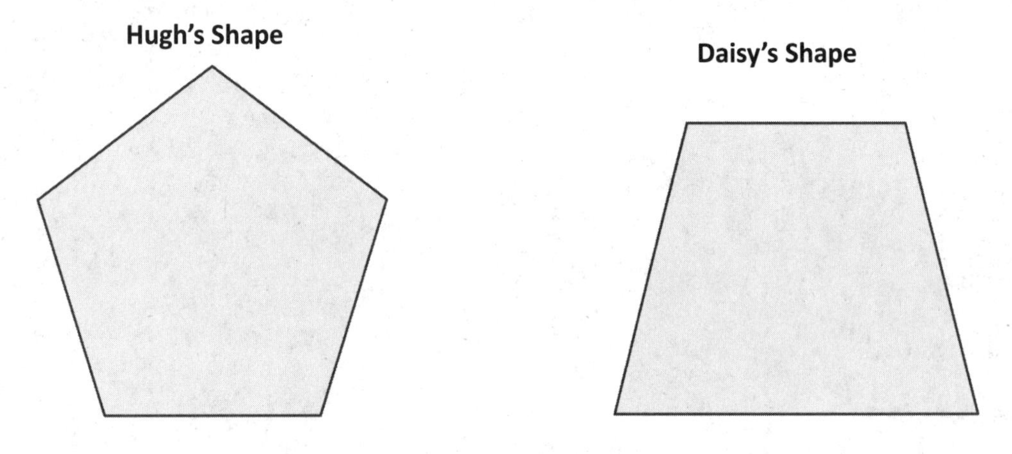

4. Andrea measures one side length of the square below and says she can find the perimeter with that measurement. Explain Andrea's thinking. Then, find the perimeter in centimeters.

Name ______________________________ Date ______________

Measure and label the side lengths of the shape below in centimeters. Then, find the perimeter.

Perimeter = ______________________________

= _______ cm

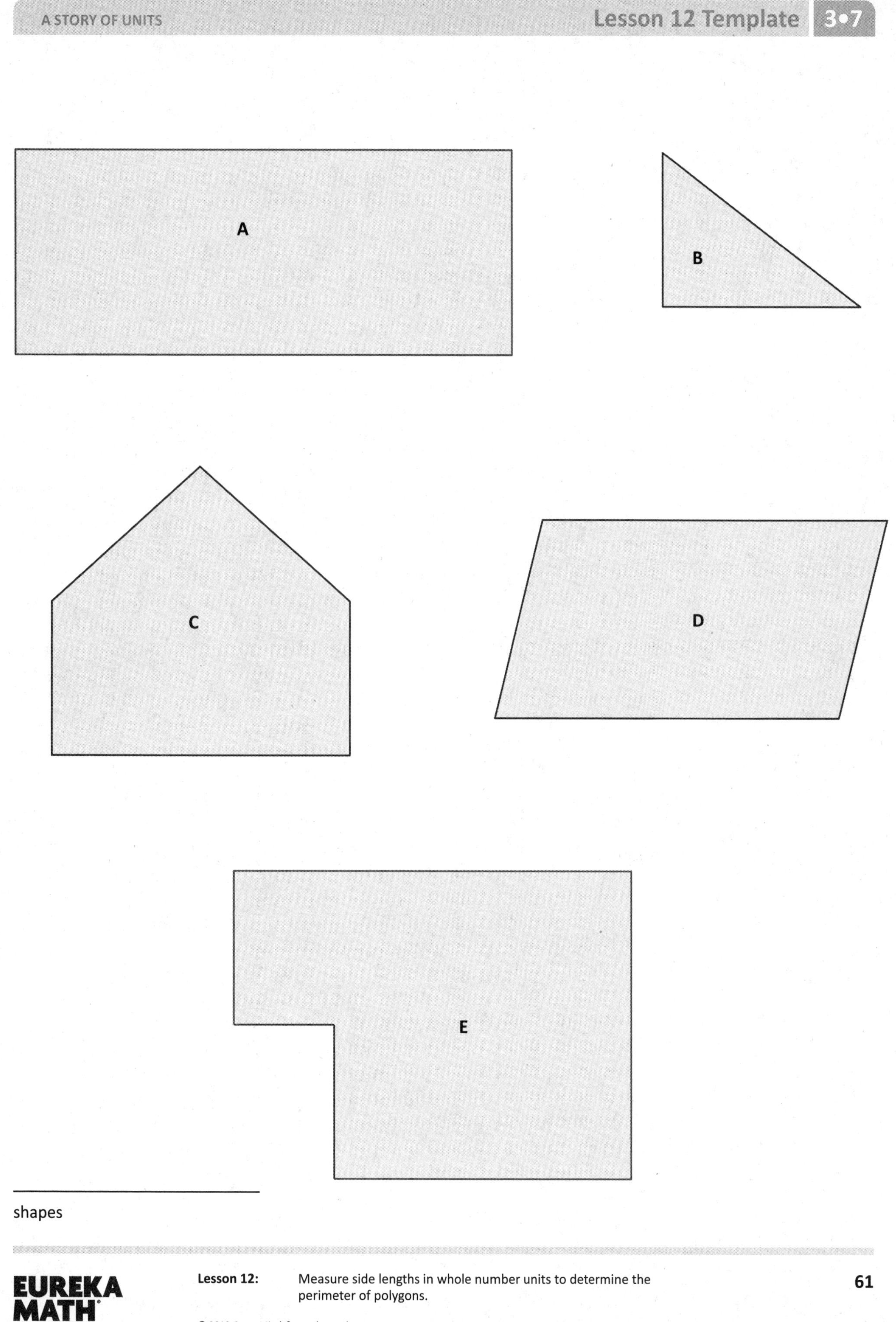

shapes

Use an index card to answer the questions.

a. What is the perimeter of your index card in inches?

b. Place the short end of your index card next to the short end of your partner's index card. Make a prediction: What do you think the perimeter is of the new shape you made?

Draw **Write**

c. Find the perimeter of the new shape. Was your prediction right? Why or why not?

Read **Draw** **Write**

Name ______________________________ Date ____________

1. Find the perimeter of the following shapes.

a.

8 in

3 in 3 in

8 in

P = 3 in + 8 in + 3 in + 8 in

= ________ in

b.

4 cm

4 cm 4 cm

4 cm

P = ____ cm + ____ cm + ____ cm + ____ cm

= ________ cm

c.

6 cm 11 cm

9 cm

P = ____ cm + ____ cm + ____ cm

= ________ cm

d.

5 m

7 m 9 m

15 m

P = ____ m + ____ m + ____ m + ____ m

= ________ m

e.

9 in

2 in

3 in

2 in

9 in

P = ____ in + ____ in + ____ in + ____ in + ____ in

= ________ in

2. Alan's rectangular swimming pool is 10 meters long and 16 meters wide. What is the perimeter?

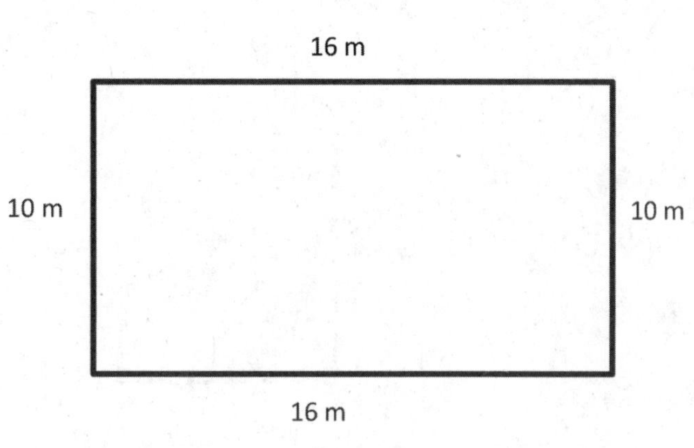

3. Lila measures each side of the shape below.

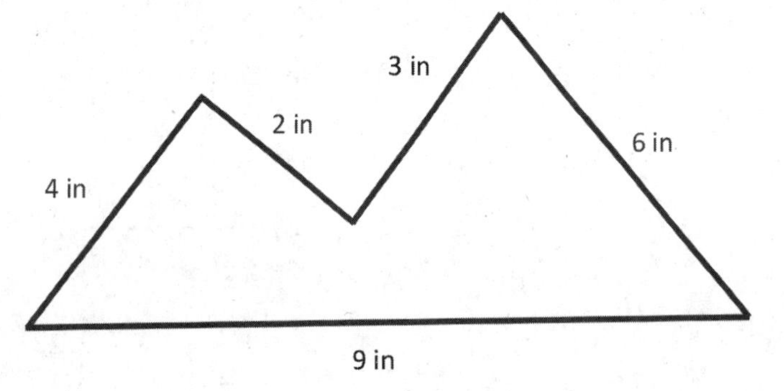

a. What is the perimeter of the shape?

b. Lila says the shape is a pentagon. Is she correct? Explain why or why not.

Name ______________________________ Date ______________

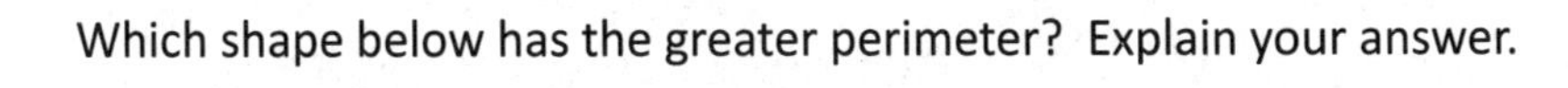

Which shape below has the greater perimeter? Explain your answer.

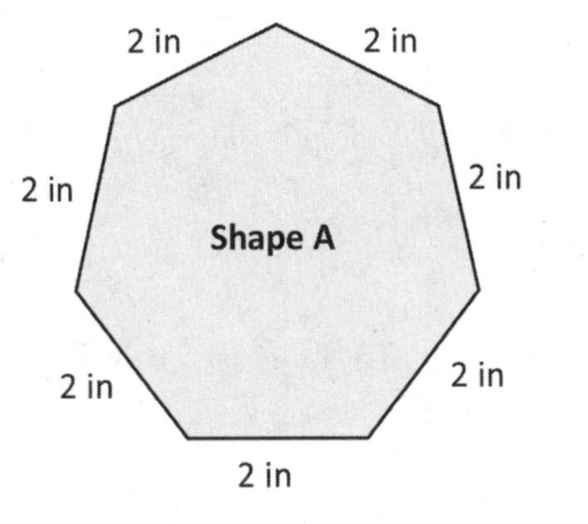

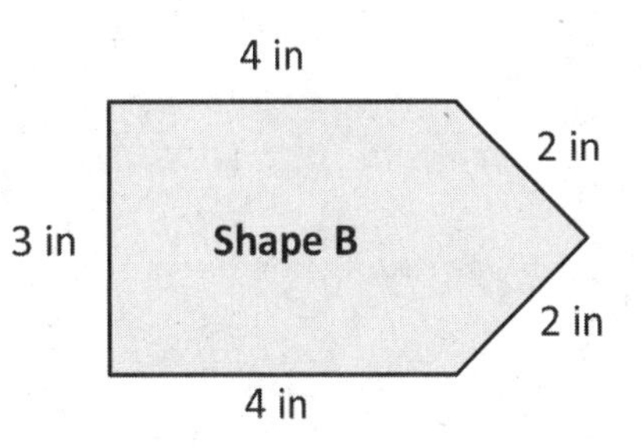

A rectangular sheep pen measures 5 meters long and 9 meters wide. The perimeter of the cow pen is double the perimeter of the sheep pen. What is the perimeter of the cow pen?

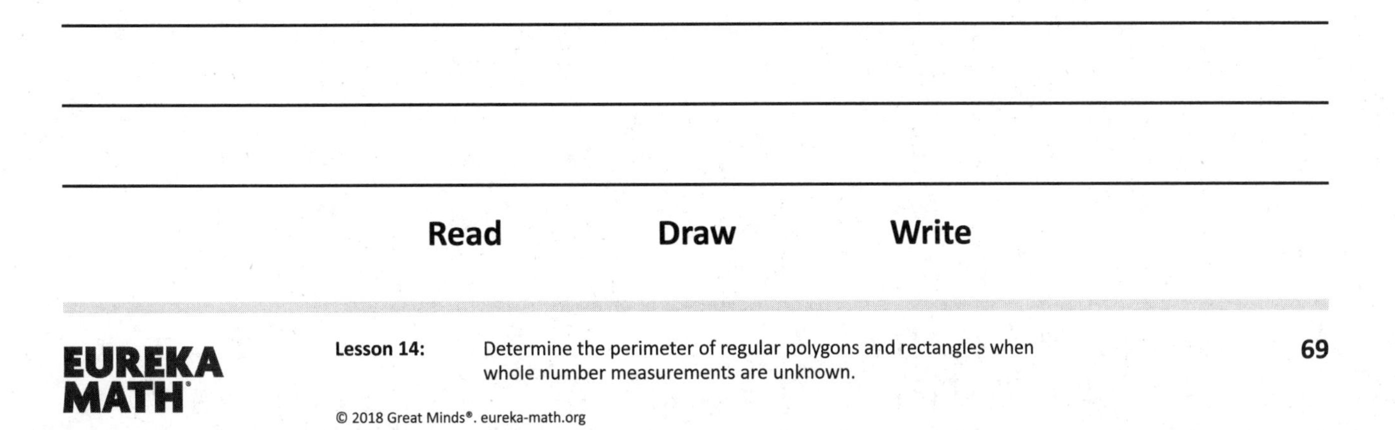

Name ______________________________ Date ______________

1. Label the unknown side lengths of the regular shapes below. Then, find the perimeter of each shape.

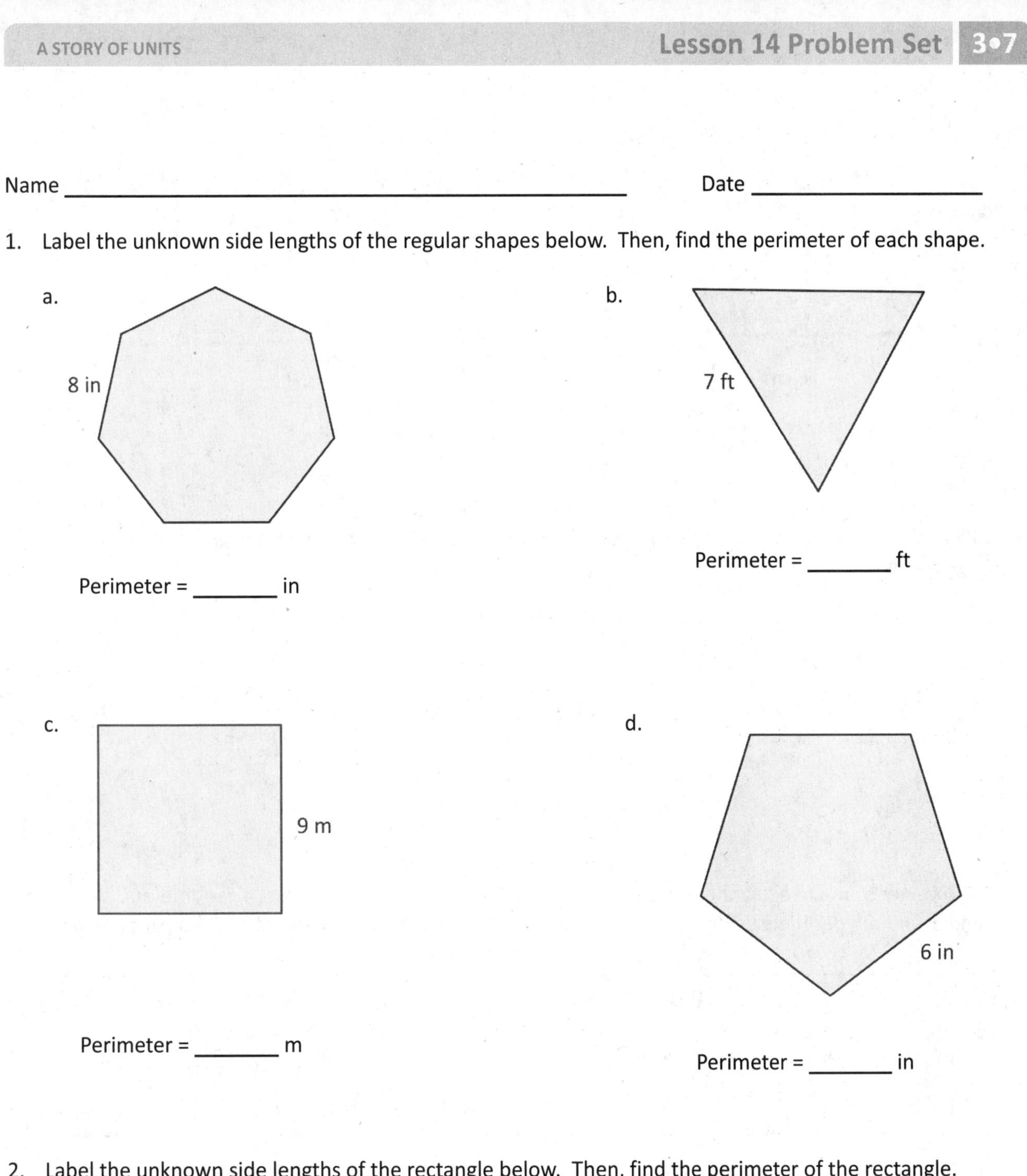

2. Label the unknown side lengths of the rectangle below. Then, find the perimeter of the rectangle.

2 cm

7 cm

Perimeter = _______ cm

3. David draws a regular octagon and labels a side length as shown below. Find the perimeter of David's octagon.

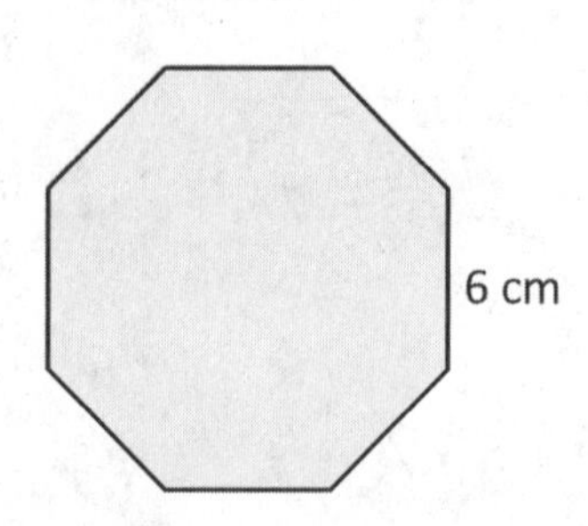

4. Paige paints an 8-inch by 9-inch picture for her mom's birthday. What is the total length of wood that Paige needs to make a frame for the picture?

5. Mr. Spooner draws a regular hexagon on the board. One of the sides measures 4 centimeters. Giles and Xander find the perimeter. Their work is shown below. Whose work is correct? Explain your answer.

Giles's Work
Perimeter = 4 cm + 4 cm + 4 cm + 4 cm + 4 cm + 4 cm
Perimeter = 24 cm

Xander's Work
Perimeter = 6 × 4 cm
Perimeter = 24 cm

Name ______________________________ Date ______________

Travis traces a regular pentagon on his paper. Each side measures 7 centimeters. He also traces a regular hexagon on his paper. Each side of the hexagon measures 5 centimeters. Which shape has a greater perimeter? Show your work.

Clara and Pedro each use four 3-inch by 5-inch cards to make the rectangles below. Whose rectangle has a greater perimeter?

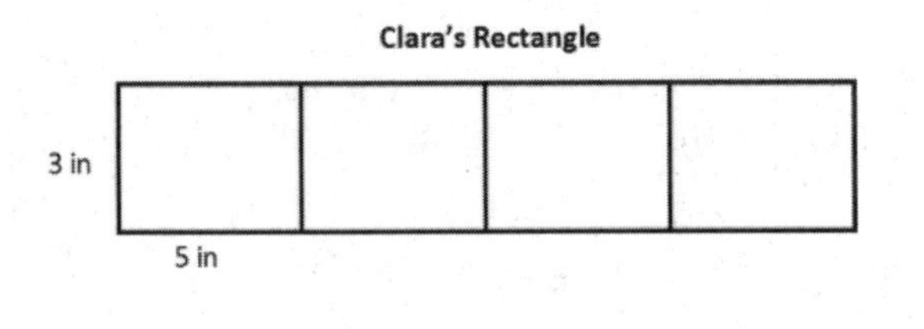

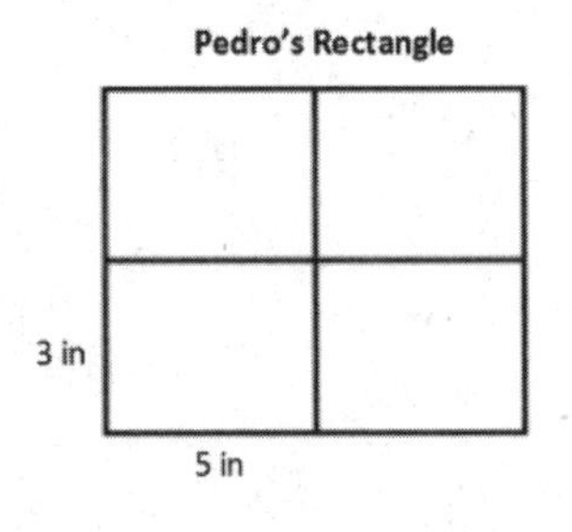

Read **Draw** **Write**

Name ______________________________ Date ______________

1. Mrs. Kozlow put a border around a 5-foot by 6-foot rectangular bulletin board. How many feet of border did Mrs. Kozlow use?

2. Jason built a model of the Pentagon for a social studies project. He made each outside wall 33 centimeters long. What is the perimeter of Jason's model pentagon?

3. The Holmes family plants a rectangular 8-yard by 9-yard vegetable garden. How many yards of fencing do they need to put a fence around the garden?

4. Marion paints a 5-pointed star on her bedroom wall. Each side of the star is 18 inches long. What is the perimeter of the star?

5. The soccer team jogs around the outside of the soccer field twice to warm up. The rectangular field measures 60 yards by 100 yards. What is the total number of yards the team jogs?

6. Troop 516 makes 3 triangular flags to carry at a parade. They sew ribbon around the outside edges of the flags. The flags' side lengths each measure 24 inches. How many inches of ribbon does the troop use?

Name ______________________________ Date ______________

Marlene ropes off a square section of her yard where she plants grass. One side length of the square measures 9 yards. What is the total length of rope Marlene uses?

Name ______________________________ Date ______________

1. Find the perimeter of 10 circular objects to the nearest quarter inch using string. Record the name and perimeter of each object in the chart below.

Object	Perimeter (to the nearest quarter inch)

a. Explain the steps you used to find the perimeter of the circular objects in the chart above.

b. Could the same process be used to find the perimeter of the shape below? Why or why not?

2. Can you find the perimeter of the shape below using just your ruler? Explain your answer.

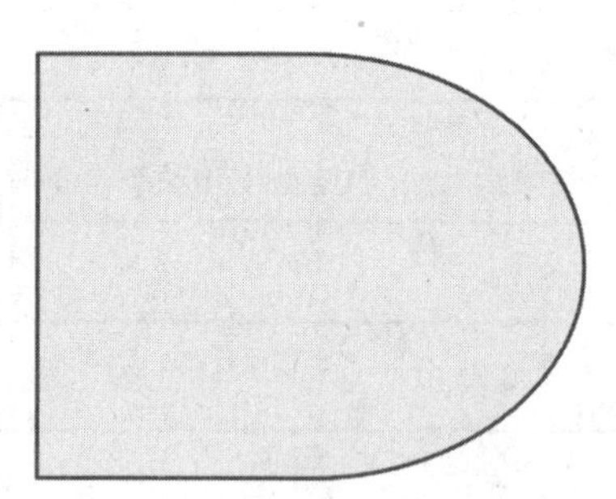

3. Molly says the perimeter of the shape below is $6\frac{1}{4}$ inches. Use your string to check her work. Do you agree with her? Why or why not?

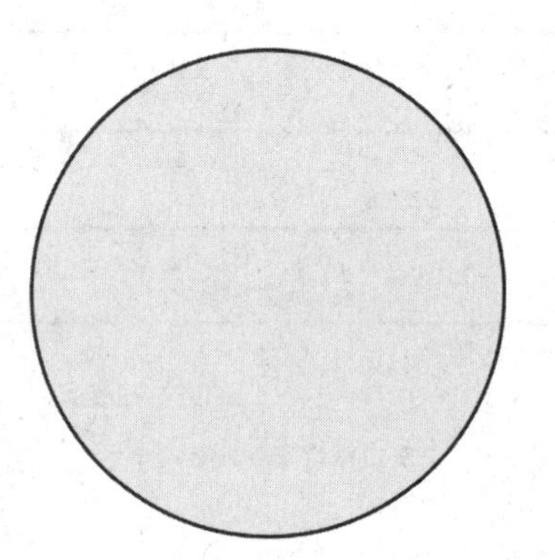

4. Is the process you used to find the perimeter of a circular object an efficient method to find the perimeter of a rectangle? Why or why not?

Name ______________________________ Date ________________

Use your string to the find the perimeter of the shape below to the nearest quarter inch.

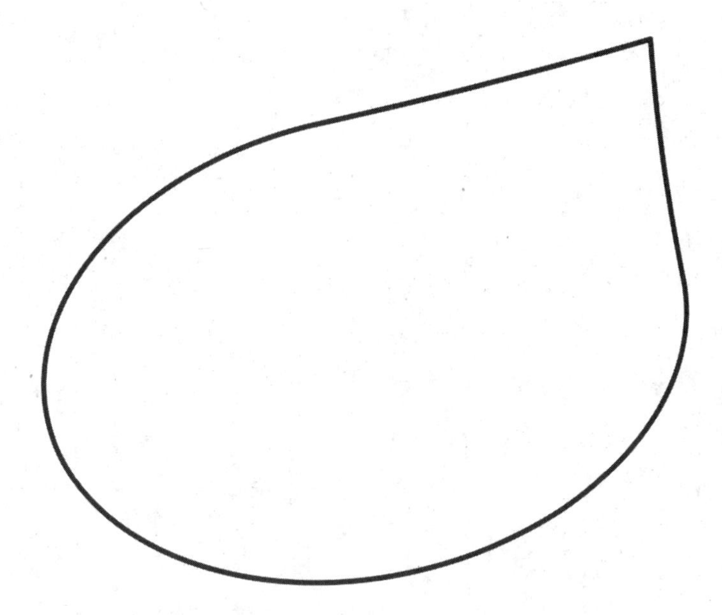

Gil places two regular hexagons side by side as shown to make a new shape. Each side measures 6 centimeters. Find the perimeter of his new shape.

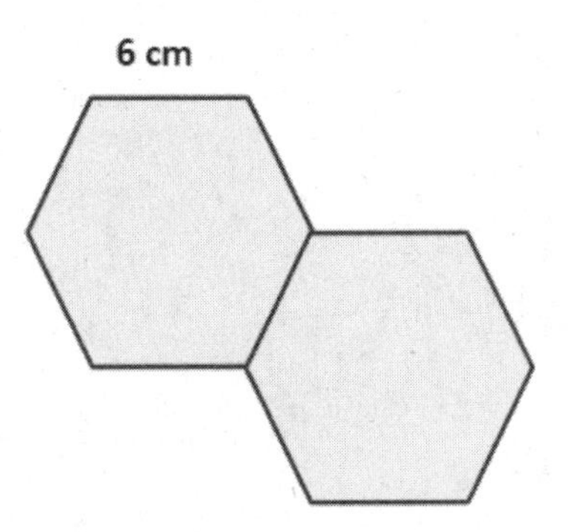

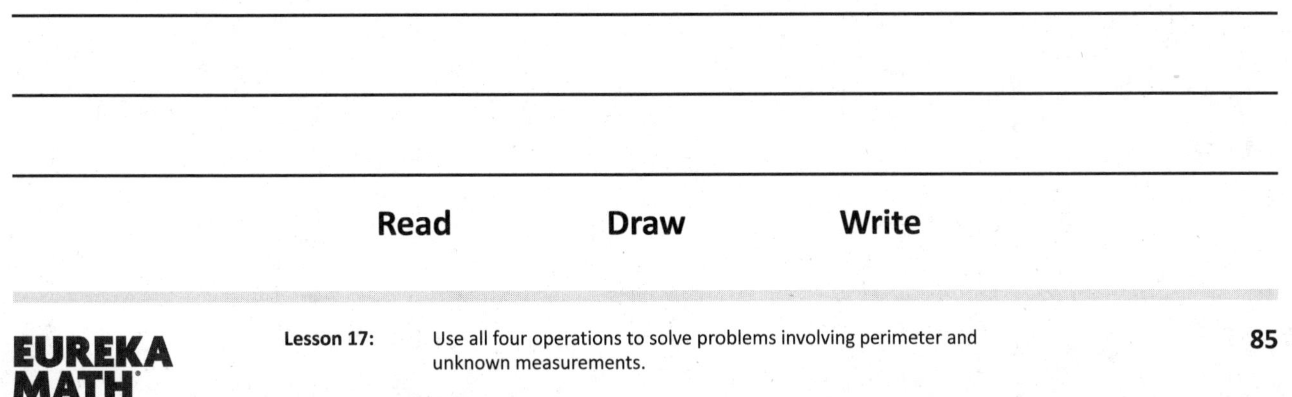

Name ______________________________ Date ____________

1. The shapes below are made up of rectangles. Label the unknown side lengths. Then, write and solve an equation to find the perimeter of each shape.

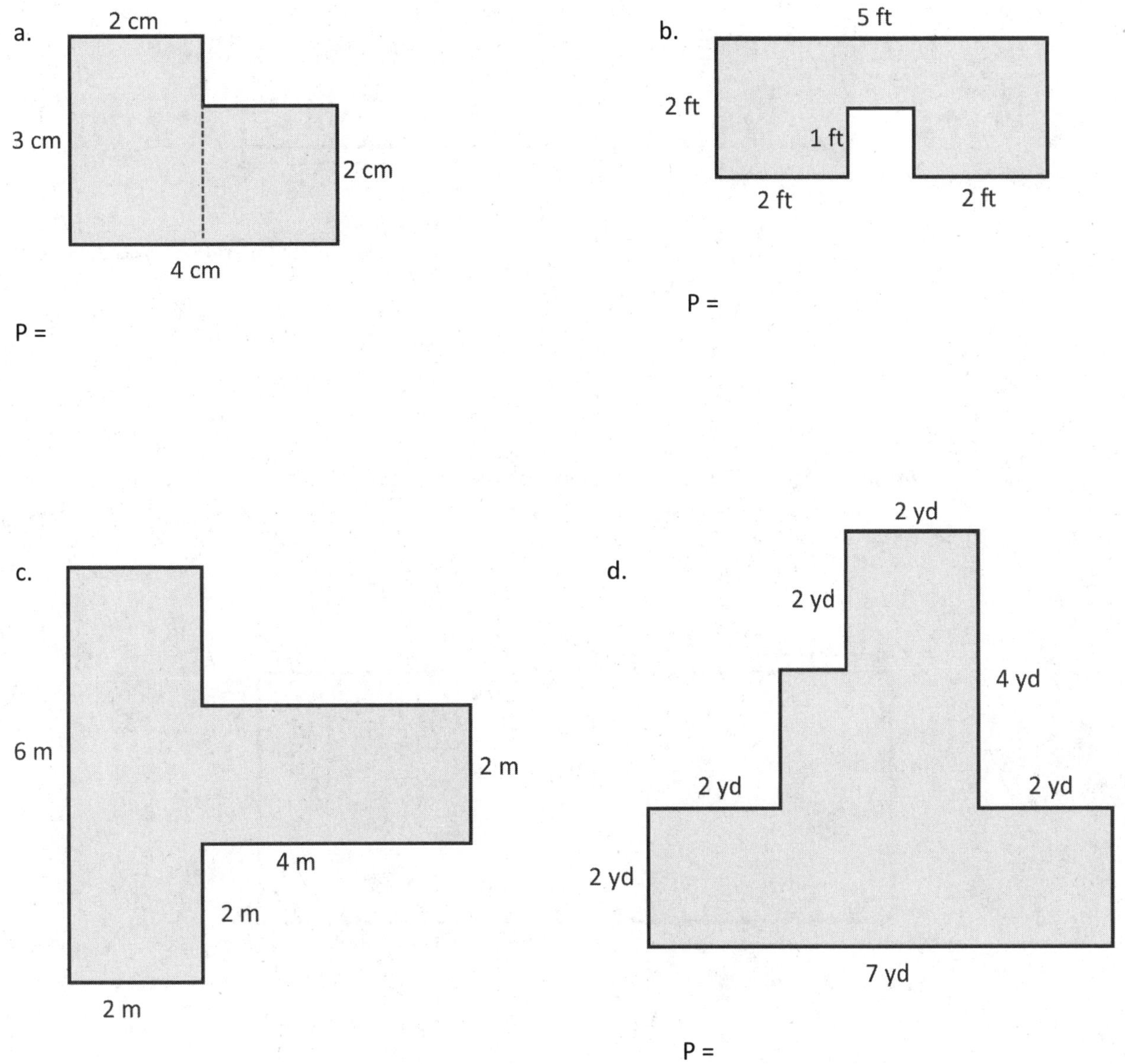

2. Nathan draws and labels the square and rectangle below. Find the perimeter of the new shape.

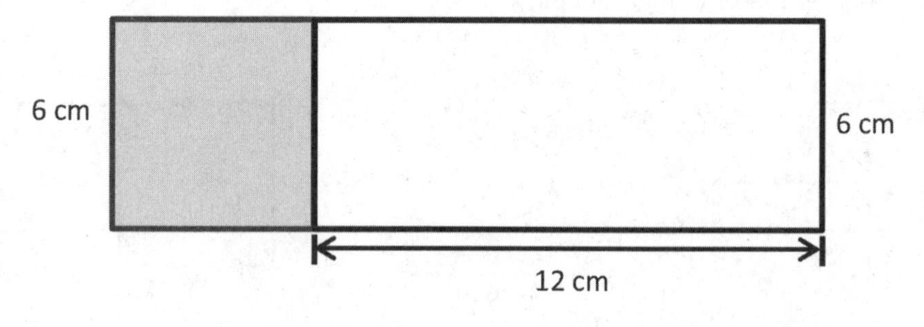

3. Label the unknown side lengths. Then, find the perimeter of the shaded rectangle.

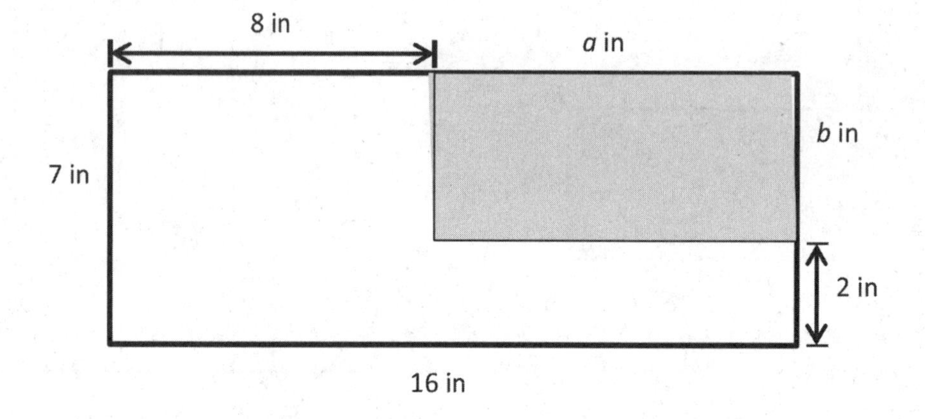

Name ______________________________ Date ______________

Label the unknown side lengths. Then, find the perimeter of the shaded rectangle.

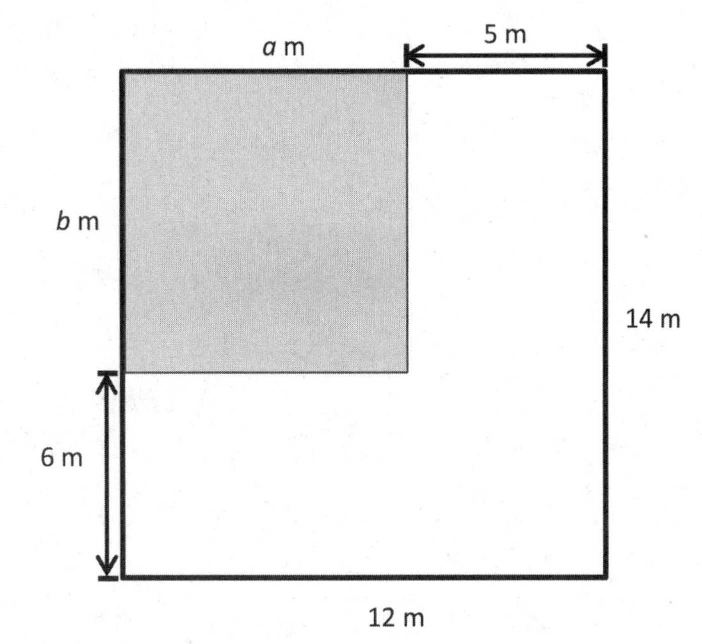

Rita says that since 15 is larger than 12, she can draw more arrays to show 15 than she can to show 12. Is she correct? Model to solve.

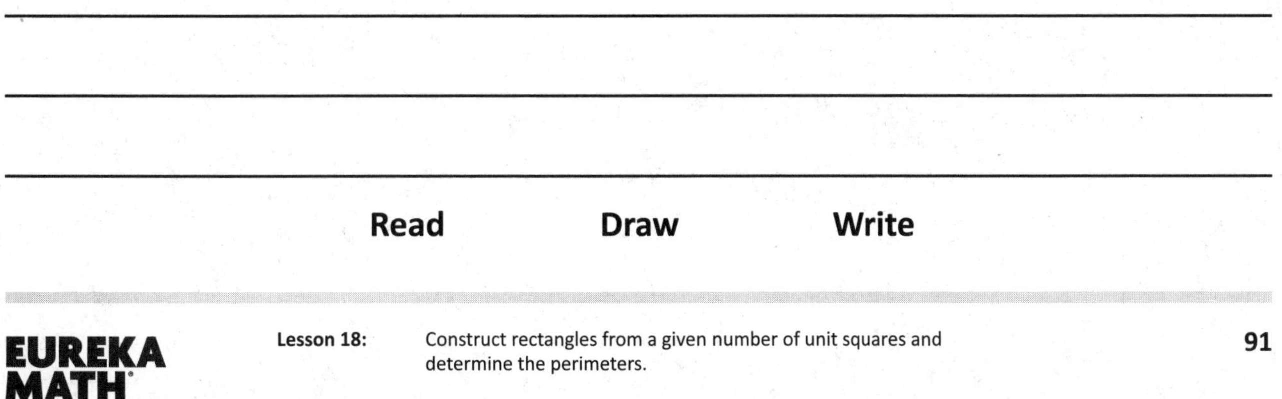

Name ___________________________________ Date ________________

1. Use unit squares to build as many rectangles as you can with an area of 24 square units. Shade in squares on your grid paper to represent each rectangle that you made with an area of 24 square units.

   a. Estimate to draw and label the side lengths of each rectangle you built in Problem 1. Then, find the perimeter of each rectangle. One rectangle is done for you.

24 units

1 unit

P = 24 units + 1 unit + 24 units + 1 unit = 50 units

   b. The areas of the rectangles in part (a) above are all the same. What do you notice about the perimeters?

2. Use unit square tiles to build as many rectangles as you can with an area of 16 square units. Estimate to draw each rectangle below. Label the side lengths.

   a. Find the perimeters of the rectangles you built.

   b. What is the perimeter of the square? Explain how you found your answer.

3. Doug uses square unit tiles to build rectangles with an area of 15 square units. He draws the rectangles as shown below but forgets to label the side lengths. Doug says that Rectangle A has a greater perimeter than Rectangle B. Do you agree? Why or why not?

Rectangle A

Rectangle B

Name ______________________________ Date ____________

Tessa uses square-centimeter tiles to build rectangles with an area of 12 square centimeters. She draws the rectangles as shown below. Label the unknown side lengths of each rectangle. Then, find the perimeter of each rectangle.

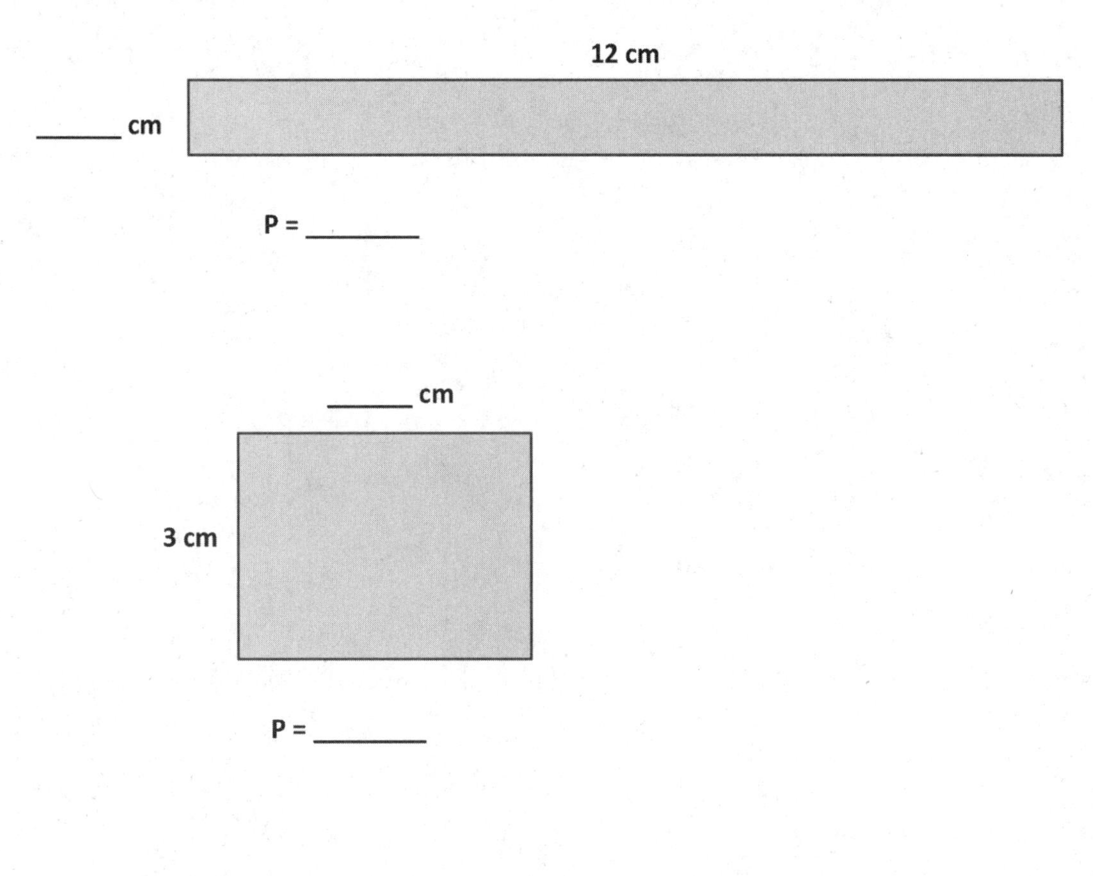

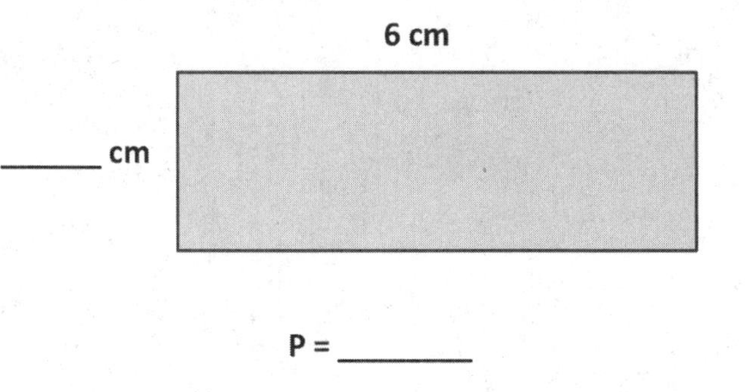

P = _______

grid paper

Marci says, "If a rectangle has a greater area than another rectangle, it must have a larger perimeter." Do you agree or disagree? Show an example to prove your thinking.

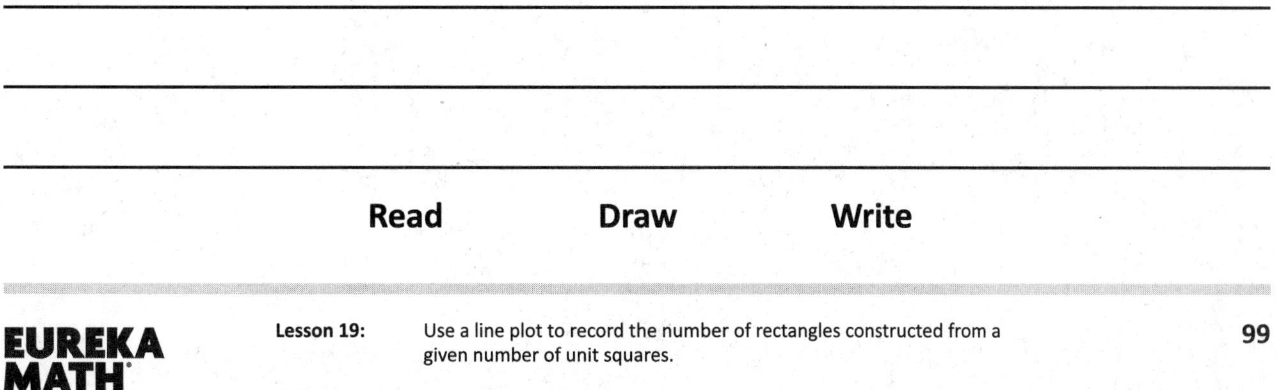

Name ______________________________ Date ______________

1. Use unit square tiles to make rectangles for each given number of unit squares. Complete the charts to show how many rectangles you can make for each given number of unit squares. The first one is done for you. You might not use all the spaces in each chart.

Number of unit squares = **12**

Number of rectangles I made: 3

Width	Length
1	12
2	6
3	4

Number of unit squares = **13**

Number of rectangles I made: ____

Width	Length

Number of unit squares = **14**

Number of rectangles I made: ____

Width	Length

Number of unit squares = **15**

Number of rectangles I made: ____

Width	Length

Number of unit squares = **16**

Number of rectangles I made: ____

Width	Length

Number of unit squares = **17**

Number of rectangles I made: ____

Width	Length

Number of unit squares = **18**

Number of rectangles I made: ____

Width	Length

2. Create a line plot with the data you collected in Problem 1.

Number of Rectangles Made with Unit Squares

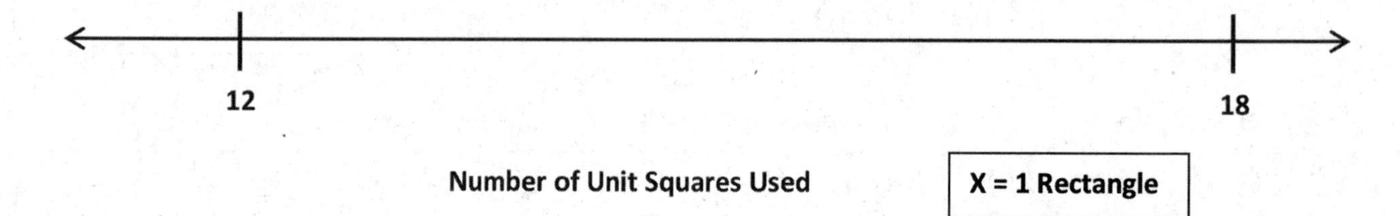

Number of Unit Squares Used

X = 1 Rectangle

3. Which numbers of unit squares produce three rectangles?

4. Why do some numbers of unit squares, such as 13, only produce one rectangle?

Name ______________________________ Date ________________

Use unit square tiles to make rectangles for the given number of unit squares. Complete the chart to show how many rectangles you made for the given number of unit squares. You might not use all the spaces in the chart.

Number of unit squares = **20** Number of rectangles I made: ____	
Width	Length

Molly builds a rectangular playpen for her pet rabbit. The playpen has an area of 15 square yards.

a. Estimate to draw and label as many possibilities as you can for the playpen.

b. Find the perimeters of the rectangles in part (a).

Read **Draw** **Write**

c. What other information do you need in order to re-create Molly's playpen?

Read **Draw** **Write**

Name ______________________________ Date ____________

1. Use your square unit tiles to build as many rectangles as you can with a perimeter of 12 units.

   a. Estimate to draw your rectangles below. Label the side lengths of each rectangle.

   b. Explain your strategy for finding rectangles with a perimeter of 12 units.

   c. Find the areas of all the rectangles in part (a) above.

   d. The perimeters of all the rectangles are the same. What do you notice about their areas?

2. Use your square unit tiles to build as many rectangles as you can with a perimeter of 14 units.

   a. Estimate to draw your rectangles below. Label the side lengths of each rectangle.

   b. Find the areas of all the rectangles in part (a) above.

   c. Given a rectangle's perimeter, what other information do you need to know about the rectangle to find its area?

Name ______________________________________ Date ________________

Use your square unit tiles to build as many rectangles as you can with a perimeter of 8 units.

a. Estimate to draw your rectangles below. Label the side lengths of each rectangle.

b. Find the areas of the rectangles in part (a) above.

Name ______________________________ Date ______________

Use the data you gathered from Problem Sets 20 and 21 to complete the charts to show how many rectangles you can create with a given perimeter. You might not use all the spaces in the charts.

Perimeter = 10 units

Number of rectangles you made: _____

Width	Length	Area
1 unit	4 units	4 square units

Perimeter = 12 units

Number of rectangles you made: _____

Width	Length	Area

Perimeter = 14 units

Number of rectangles you made: _____

Width	Length	Area

Perimeter = 16 units

Number of rectangles you made: _____

Width	Length	Area

Perimeter = 18 units

Number of rectangles you made: _____

Width	Length	Area

Perimeter = 20 units

Number of rectangles you made: _____

Width	Length	Area

Mrs. Zeck will use 14 feet of tape to mark a rectangle on the gym wall. Draw several rectangles that Mrs. Zeck could make with her tape. Label the width and length of each rectangle.

Draw **Write**

Name ______________________________ Date ________________

1. On your centimeter grid paper, shade and label as many rectangles as you can with a perimeter of 16 centimeters.

   a. Sketch the rectangles below, and label the side lengths.

   b. Find the area of each rectangle you drew above.

2. On your centimeter grid paper, shade and label as many rectangles as you can with a perimeter of 18 centimeters.

   a. Sketch the rectangles below, and label the side lengths.

   b. Find the area of each rectangle you drew above.

3. Use centimeter grid paper to shade in as many rectangles as you can with the given perimeters.

   a. Use the charts below to show how many rectangles you shaded for each given perimeter. You might not use all the spaces in the charts.

Perimeter = 10 cm Number of rectangles I made: ____		
Width	Length	Area
1 cm	4 cm	4 square cm

Perimeter = 20 cm Number of rectangles I made: ____		
Width	Length	Area
1 cm	9 cm	9 square cm

   b. Did you make a square with either of the given perimeters? How do you know?

4. Macy and Gavin both draw rectangles with perimeters of 16 centimeters. Use words and pictures to explain how it is possible for Macy's and Gavin's rectangles to have the same perimeters but different areas.

Name ______________________________ Date ______________

On the grid below, shade and label at least two different rectangles with a perimeter of 20 centimeters.

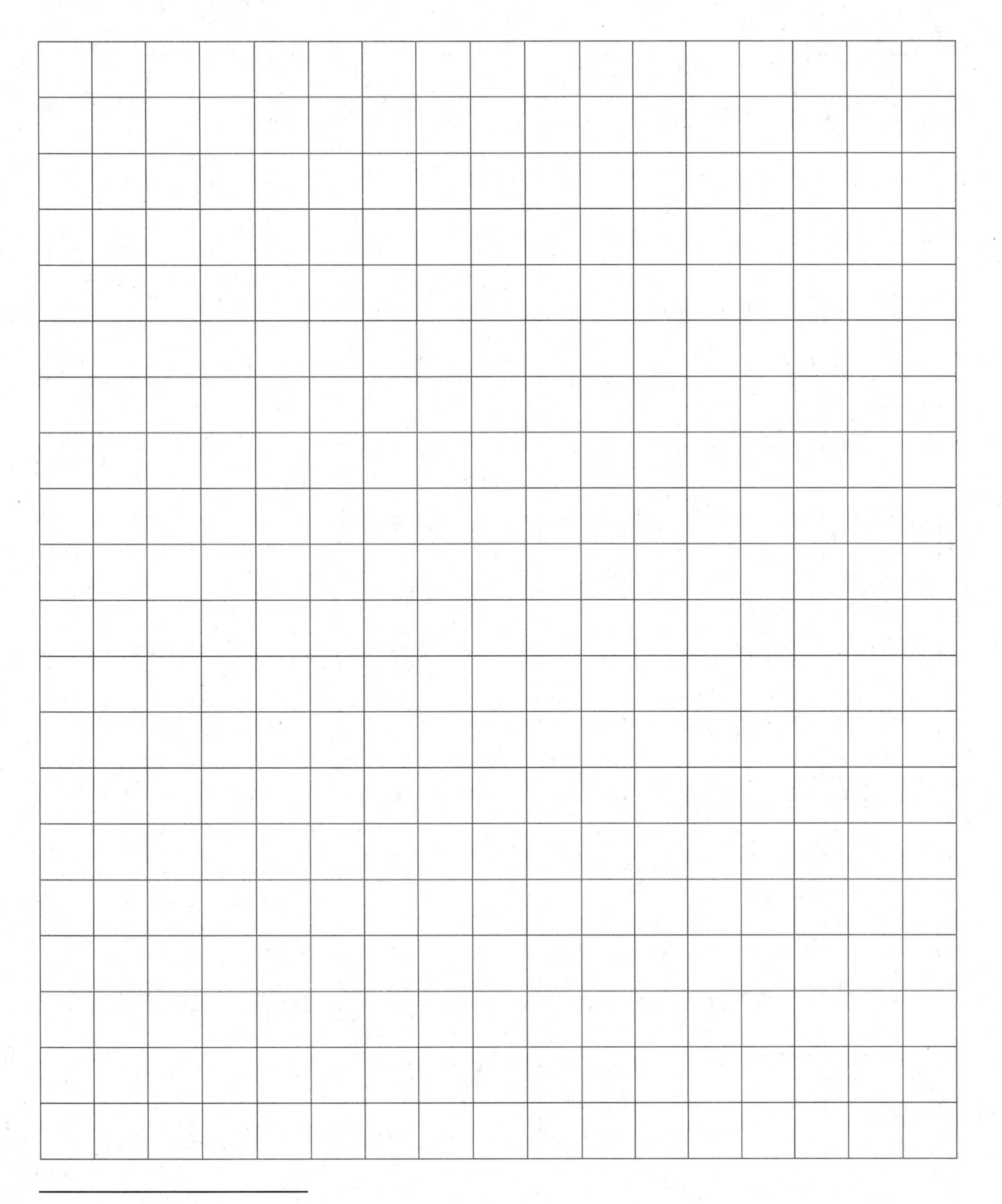

centimeter grid paper

Name ______________________________ Date ______________

Use the data you gathered from Problem Sets 20 and 21 to complete the charts to show how many rectangles you can create with a given perimeter. You might not use all the spaces in the charts.

Perimeter = 10 units

Number of rectangles you made: _____

Width	Length	Area
1 unit	4 units	4 square units

Perimeter = 12 units

Number of rectangles you made: _____

Width	Length	Area

Perimeter = 14 units

Number of rectangles you made: _____

Width	Length	Area

Perimeter = 16 units

Number of rectangles you made: _____

Width	Length	Area

Perimeter = 18 units

Number of rectangles you made: _____

Width	Length	Area

Perimeter = 20 units

Number of rectangles you made: _____

Width	Length	Area

Name ______________________________ Date ______________

1. Use the data you gathered from your Problem Sets to create a line plot for the number of rectangles you created with each given perimeter.

Number of Rectangles Made with a Given Perimeter

Perimeter Measurements in Units

X = 1 Rectangle

2. Why are all of the perimeter measurements even? Do all rectangles have an even perimeter?

3. Compare the two line plots we created. Is there any reason to think that knowing only the area of a rectangle would help you to figure out its perimeter or knowing only the perimeter of a rectangle would help you figure out its area?

4. Sumi uses unit square tiles to build 3 rectangles that have an area of 32 square units. Does knowing this help her find the number of rectangles she can build for a perimeter of 32 units? Why or why not?

5. George draws 3 rectangles that have a perimeter of 14 centimeters. Alicia tells George that there are more than 3 rectangles that have a perimeter of 14 centimeters. Explain why Alicia is correct.

Name ___________________________________ Date ________________

Suppose you have a rectangle with a perimeter of 2 cm. What can you conclude about the side lengths? Can all 4 sides of the rectangle measure a whole number of centimeters?

Name ______________________________ Date ______________

1. Gale makes a miniature stop sign, a regular octagon, with a perimeter of 48 centimeters for the town he built with blocks. What is the length of each side of the stop sign?

2. Travis bends wire to make rectangles. Each rectangle measures 34 inches by 12 inches. What is the total length of the wire needed for two rectangles?

3. The perimeter of a rectangular bathroom is 32 feet. The width of the room is 8 feet. What is the length of the room?

4. Raj uses 6-inch square tiles to make a rectangle, as shown below. What is the perimeter of the rectangle in inches?

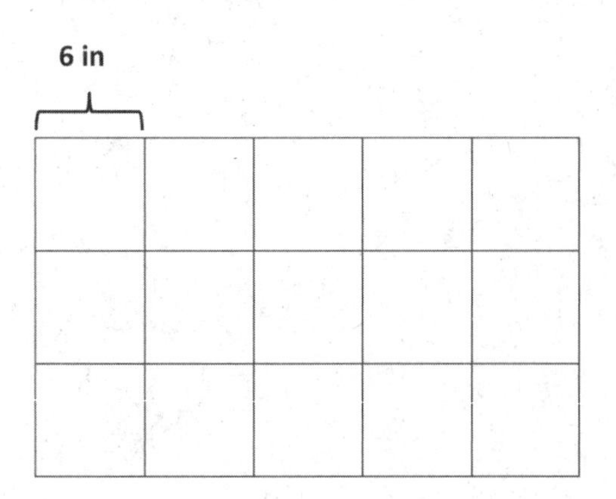

5. Mischa makes a 4-foot by 6-foot rectangular banner. She puts ribbon around the outside edges. The ribbon costs $2 per foot. What is the total cost of the ribbon?

6. Colton buys a roll of wire fencing that is 120 yards long. He uses it to fence in his 18-yard by 24-yard rectangular garden. Will Colton have enough wire fencing left over to fence in a 6-yard by 8-yard rectangular play space for his pet rabbit?

Name ______________________________ Date ______________

Adriana traces a regular triangle to create the shape below. The perimeter of her shape is 72 centimeters. What are the side lengths of the triangle?

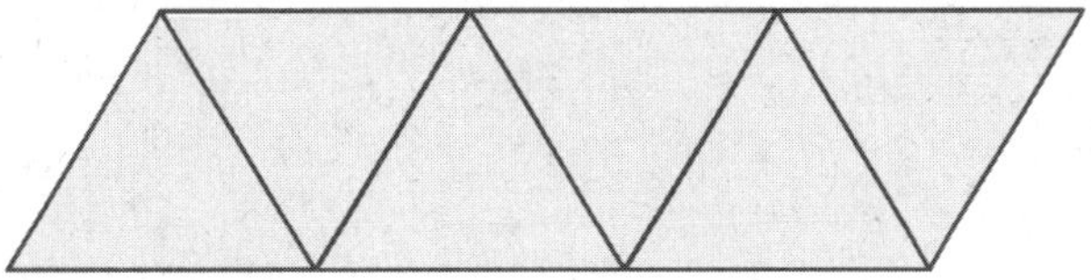

Name ________________________________ Date ______________

Use the given perimeters in the chart below to choose the widths and lengths of your robot's rectangular body parts. Write the widths and lengths in the chart below. Use the blank rows if you want to add extra rectangular body parts to your robot.

Letter	Body Part	Perimeter	Width and Length
A	arm	14 cm	________ cm by ________ cm
B	arm	14 cm	________ cm by ________ cm
C	leg	18 cm	________ cm by ________ cm
D	leg	18 cm	________ cm by ________ cm
E	body	Double the perimeter of one arm = ________ cm	________ cm by ________ cm
F	head	16 cm	________ cm by ________ cm
G	neck	Half the perimeter of the head = ________ cm	________ cm by ________ cm
H			________ cm by ________ cm
I			________ cm by ________ cm
My robot has 7 to 9 rectangular body parts. Number of body parts: ________			

Use the information in the chart below to plan an environment for your robot. Write the width and length for each rectangular item. Use the blank rows if you want to add extra circular or rectangular items to your robot's environment.

Letter	Item	Shape	Perimeter	Width and Length
J	sun	circle	about 25 cm	
K	house	rectangle	82 cm	_______ cm by _______ cm
L	tree top	circle	about 30 cm	
M	tree trunk	rectangle	30 cm	_______ cm by _______ cm
N	tree top	circle	about 20 cm	
O	tree trunk	rectangle	20 cm	_______ cm by _______ cm
P				
Q				
My robot's environment has 6 to 8 items. Number of items: _______				

Name ______________________________ Date ________________

Estimate to draw three different rectangles with a perimeter of 16 centimeters. Label the width and length of each rectangle.

Name ____________________ Date ____________

Draw a picture of your robot in its environment in the space below. Label the widths, lengths, and perimeters of all rectangles. Label the perimeters of all circular shapes.

Name ______________________________ Date ______________

1. Sketch rectangles with the following perimeters. Label the side lengths.

   a. 22 cm

   b. 30 cm

2. Explain the steps you took to create the rectangles with the given perimeters.

Drew makes rectangular shoes for his robot. Each shoe has whole number side lengths and an area of 7 square centimeters. What is the total perimeter of both shoes? Is there more than one answer? Why or why not?

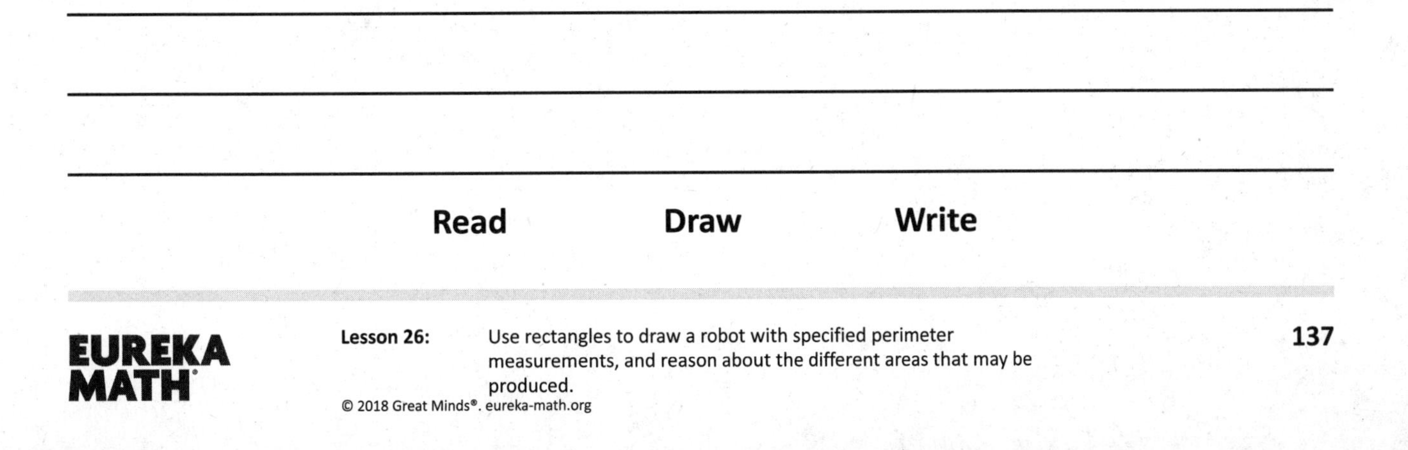

Read Draw Write

Name ____________________________________ Date ________________

1. Collect the area measurements of your classmates' **robot bodies.** Make a line plot using everyone's area measurements.

Areas of Robot Bodies

Area Measurements of the Robot's Body in Square Centimeters

X = 1 Robot Body

a. How many different measurements are on the line plot? Why are the measurements different?

b. What does this tell you about the relationship between area and perimeter?

2. Measure and calculate the perimeter of your construction paper in inches. Show your work below.

3. Sketch and label two shapes with the same perimeter from the robot's environment. What do you notice about the way they look?

4. Write two or three sentences describing your robot and the environment in which it lives.

Name ______________________________ Date ____________

1. Use string to help you sketch a circle with a perimeter of about 15 centimeters.

2. Estimate to draw a rectangle with a perimeter of 15 centimeters. Label the width and length.

Name ______________________________ Date ______________

Part A: I reviewed ______________ 's robot.

1. Use the chart below to evaluate your friend's robot. Measure the width and length of each rectangle. Then, calculate the perimeter. Record that information in the chart below. If your measurements differ from those listed on the project, put a star by the letter of the rectangle.

Rectangle	Width and Length	Student's Perimeter	Required Perimeter
A	________ cm by ________ cm		14 cm
B	________ cm by ________ cm		14 cm
C	________ cm by ________ cm		18 cm
D	________ cm by ________ cm		18 cm
E	________ cm by ________ cm		28 cm
F	________ cm by ________ cm		16 cm
G	________ cm by ________ cm		8 cm
H	________ cm by ________ cm		
I	________ cm by ________ cm		

2. Is the perimeter of the robot's body double that of the arm? Show calculations below.

3. Is the perimeter of the robot's neck half the perimeter of the head? Show calculations below.

Part B: I reviewed ________________'s robot environment.

4. Use the chart below to evaluate your friend's robot environment. Measure the width and length of each rectangle. Then, calculate the perimeter. Use your string to measure the perimeters of nonrectangular items. Record that information in the chart below. If your measurements differ from those listed on the project, put a star by the letter of the shape.

Item	Width and Length	Student's Perimeter	Required Perimeter
J			About 25 cm
K	________ cm by ________ cm		82 cm
L			About 30 cm
M	________ cm by ________ cm		30 cm
N			About 20 cm
O	________ cm by ________ cm		20 cm
P			
Q			

Name ______________________________ Date ____________

1. Record the perimeters and areas of Rectangles A and B in the chart below.

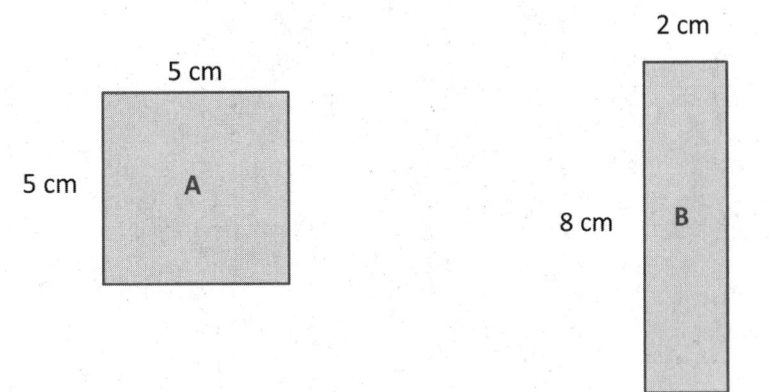

Rectangle:	Width and Length:	Perimeter	Area
A	______ cm by ______ cm		
B	______ cm by ______ cm		

2. What is the same about Rectangles A and B? What is different?

Name Sample Date

Part A: I reviewed Student A 's robot.

Use the chart below to evaluate your friend's robot. Measure the lengths and widths of each rectangle. Then calculate the perimeter. Record that information in the table below. If your measurements differ from those listed on the project, put a star by the letter of the rectangle.

Rectangle	Width and Length	Student's Perimeter	Required Perimeter
A	2 cm by 5 cm	2cm + 2cm + 5cm + 5cm = 14cm	14 cm
B	2 cm by 5 cm		14 cm
C	2 cm by 7 cm		18 cm
D	2 cm by 7 cm		18 cm
E	6 cm by 8 cm		28 cm
F	4 cm by 4 cm		16 cm
G	2 cm by 2 cm		8 cm
H	____ cm by ____ cm		
I	____ cm by ____ cm		

sample Problem Set

Name ______________________________ Date ______________

1. Gia measures her rectangular garden and finds the width is 9 yards and the length is 7 yards.

   a. Estimate to draw Gia's garden, and label the side lengths.

   b. What is the area of Gia's garden?

   c. What is the perimeter of Gia's garden?

2. Elijah draws a square that has side lengths of 8 centimeters.

   a. Estimate to draw Elijah's square, and label the side lengths.

   b. What is the area of Elijah's square?

   c. What is the perimeter of Elijah's square?

d. Elijah connects three of these squares to make one long rectangle. What is the perimeter of this rectangle?

3. The area of Mason's rectangular painting is 72 square inches. The width of the painting is 8 inches.

   a. Estimate to draw Mason's painting, and label the side lengths.

   b. What is the length of the painting?

   c. What is the perimeter of Mason's painting?

   d. Mason's mom hangs the painting on a wall that already has two of Mason's other paintings. The areas of the other paintings are 64 square inches and 81 square inches. What is the total area of the wall that is covered with Mason's paintings?

4. The perimeter of Jillian's rectangular bedroom is 34 feet. The length of her bedroom is 9 feet.

   a. Estimate to draw Jillian's bedroom, and label the side lengths.

   b. What is the width of Jillian's bedroom?

   c. What is the area of Jillian's bedroom?

   d. Jillian has a 4-foot by 6-foot rug in her room. What is the area of the floor that is not covered by the rug?

Name ____________________________________ Date ________________

Jennifer measures her rectangular sandbox and finds the width is 8 feet and the length is 6 feet.

a. Estimate to draw Jennifer's sandbox, and label the side lengths.

b. What is the area of Jennifer's sandbox?

c. What is the perimeter of Jennifer's sandbox?

Name ______________________________ Date ______________

1. Kyle puts two rectangles together to make the L-shaped figure below. He measures some of the side lengths and records them as shown.

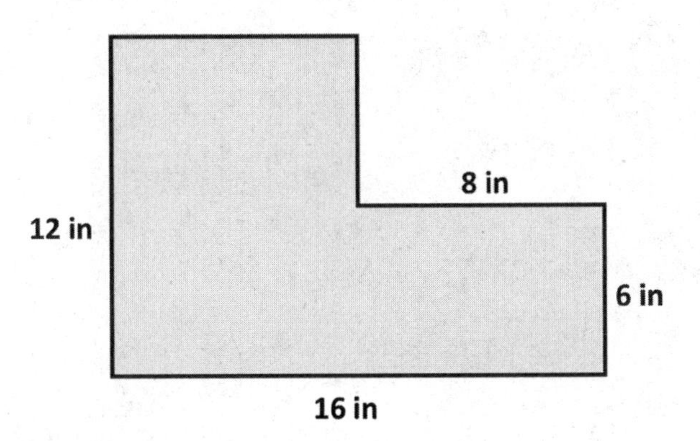

a. Find the perimeter of Kyle's shape.

b. Find the area of Kyle's shape.

c. Kyle makes two copies of the L-shaped figure to create the rectangle shown below. Find the perimeter of the rectangle.

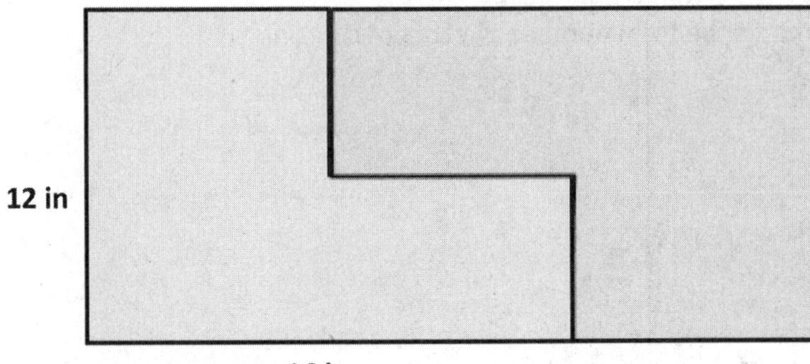

2. Jeremiah and Hayley use a piece of rope to mark a square space for their booth at the science fair. The area of their space is 49 square feet. What is the length of the rope that Jeremiah and Hayley use if they leave a 3-foot opening so they can get in and out of the space?

3. Vivienne draws four identical rectangles as shown below to make a new, larger rectangle. The perimeter of one of the small rectangles is 18 centimeters, and the width is 6 centimeters. What is the perimeter of the new, larger rectangle?

4. A jogging path around the outside edges of a rectangular playground measures 48 yards by 52 yards. Maya runs $3\frac{1}{2}$ laps on the jogging path. What is the total number of yards Maya runs?

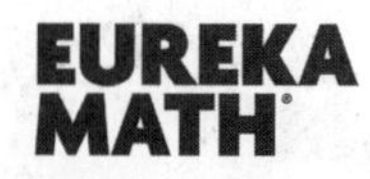

Name ____ Date __________________

Jeannette draws four identical squares as shown below to make a new, larger square. The length of one of the small square sides is 8 centimeters. What is the perimeter of the new, larger square?

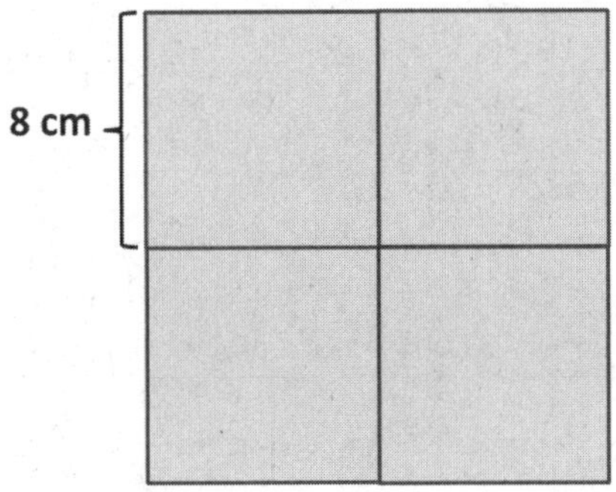

8 cm

Name ____________________________________ Date ________________

Use this form to critique your classmate's problem-solving work.

Classmate:		**Problem Number:**	
Strategies My Classmate Used:			
Things My Classmate Did Well:			
Suggestions for Improvement:			
Strategies I Would Like to Try Based on My Classmate's Work:			

Name ______________________________ Date ______________

Jayden solves the problem as shown below.

The recreation center soccer field measures 35 yards by 65 yards. Chris dribbles the soccer ball around the field 4 times. What is the total number of yards Chris dribbles the ball?

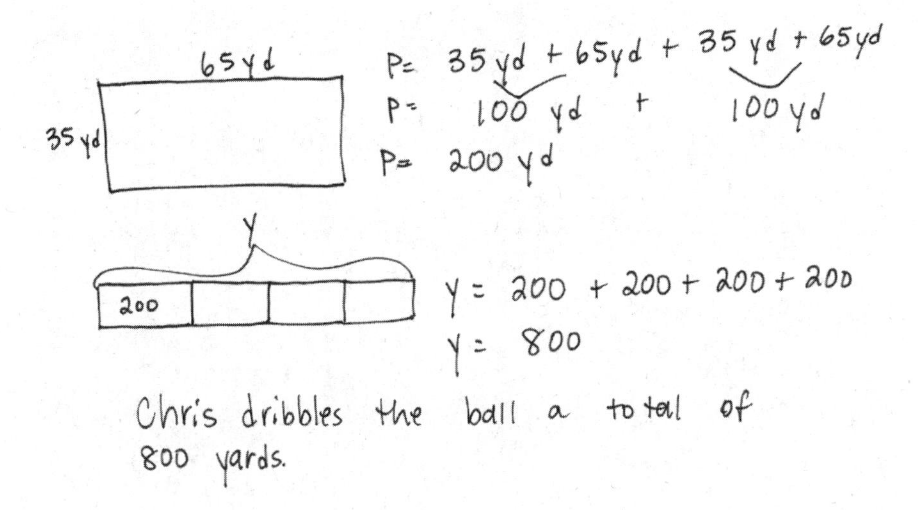

1. What strategies did Jayden use to solve this problem?

2. What did Jayden do well?

Student A

7 ft

7 ft | A = 49 sq ft

$7 \times 7 = 49$

$P = 7\text{ ft} + 7\text{ ft} + 7\text{ ft} + 7\text{ ft}$

$P = 4 \times 7\text{ ft}$

$P = 28\text{ ft}$

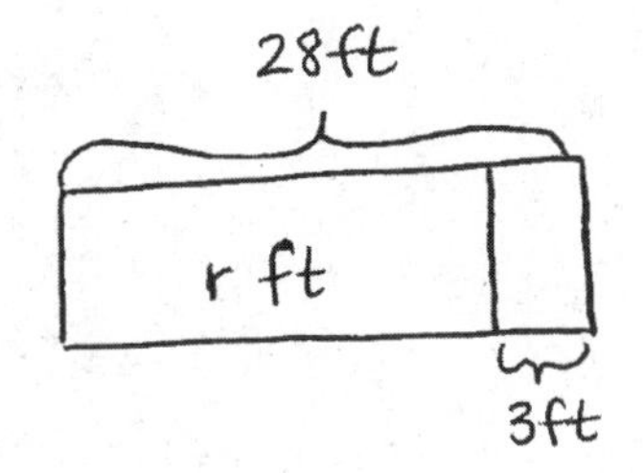

$r = 28 - 3$

$r = 25$

The total length of the rope is 25 feet.

Student B

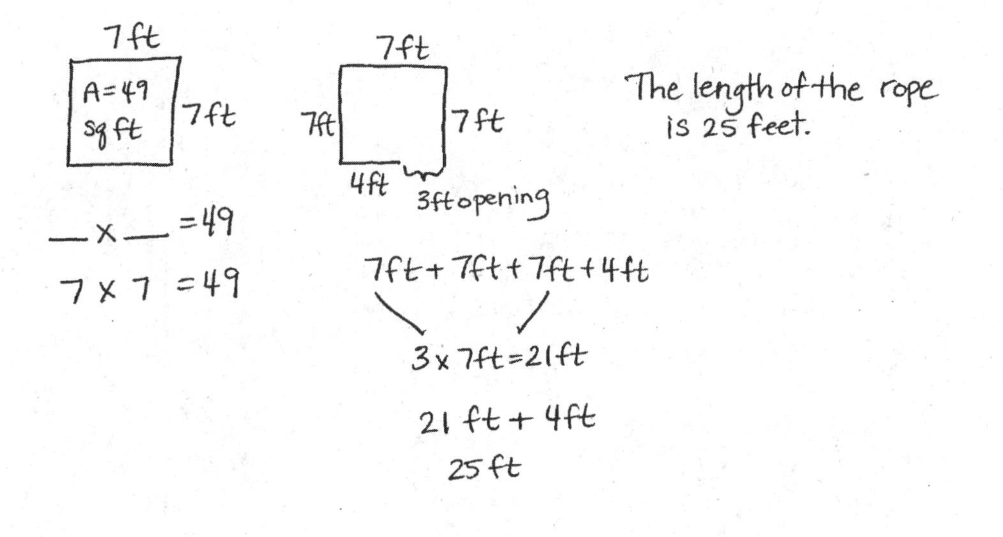

Student C

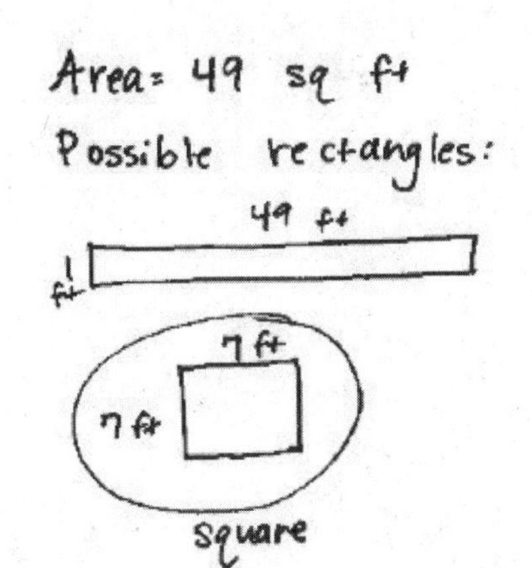

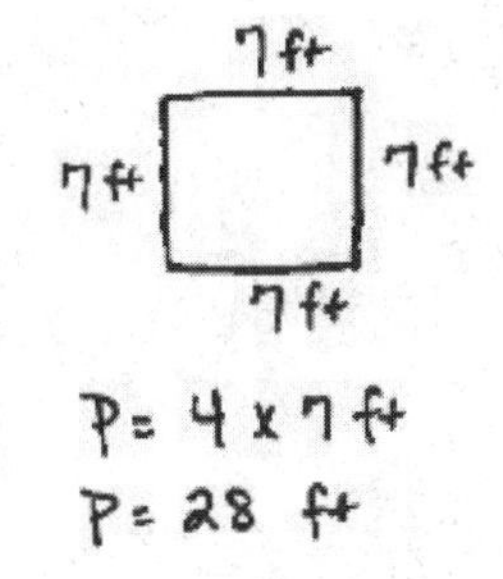

28 ft − 3 ft = 25 ft

The length of the rope is 25 ft.

student work sample images

Mara draws a 6-inch by 8-inch rectangle. She shades one-half of the rectangle. What is the area of the shaded part of Mara's rectangle?

Read **Draw** **Write**

Name ______________________________ Date ______________

Use this form to analyze your classmate's representations of one-half shaded.

Square (letter)	Does this square show one-half shaded?	Explain why or why not.	Describe changes to make so the square shows one-half shaded.

Name ____________________________________ Date ____________________

Marty shades the square as shown below and says one-half of the big square is shaded. Do you agree? Why or why not?

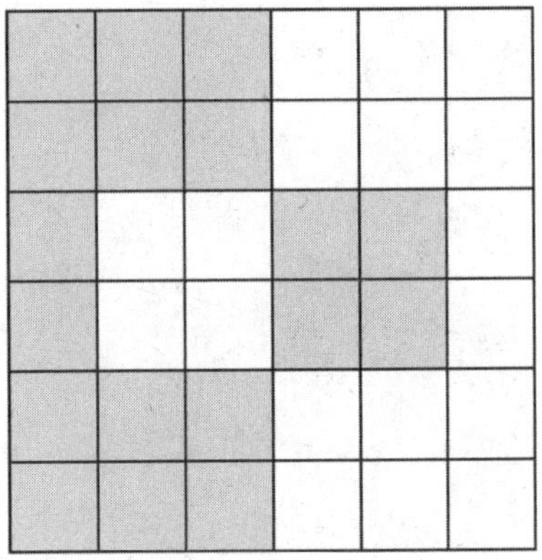

squares

Hannah traces square-inch tiles to draw 3 larger squares. She draws the 3 large squares side by side to make a rectangle. She shades one-half of each larger square, as shown.

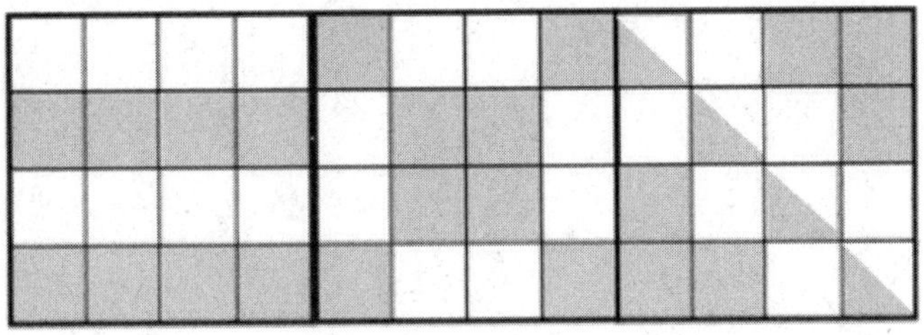

a. Do you agree that all 3 squares are one-half shaded? Explain your answer.

b. What is the area of the rectangle?

Read **Draw** **Write**

c. What is the total area of the shaded space?

Read **Draw** **Write**

Name ______________________________ Date ________________

1. Look at the circles you shaded today. Glue a circle that is about one-half shaded in the space below.

   a. Explain the strategy you used to shade in one-half of your circle.

   b. Is your circle exactly one-half shaded? Explain your answer.

2. Julian shades 4 circles as shown below.

Circle A

Circle B

Circle C

Circle D

   a. Write the letters of the circles that are about one-half shaded.

b. Choose one circle from your answer to Part (a), and explain how you know it's about one-half shaded.

Circle _______

c. Choose one circle that you did not list in Part (a), and explain how it could be changed so that it is about one-half shaded.

Circle _______

3. Read the clues to help you shade the circle below.

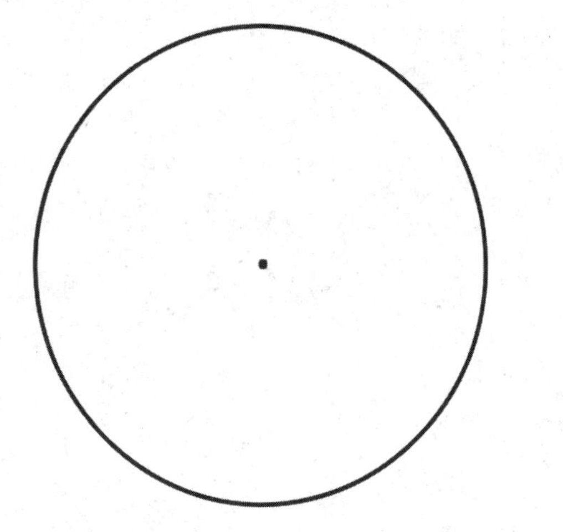

a. Divide the circle into 4 equal parts.

b. Shade in 2 parts.

c. Erase a small circle from each shaded part.

d. Estimate to draw and shade 2 circles in the unshaded parts that are the same size as the circles you erased in Part (c).

4. Did you shade in one-half of the circle in Problem 3? How do you know?

Name ______________________________ Date ______________

Riddian shades a circle as shown below.

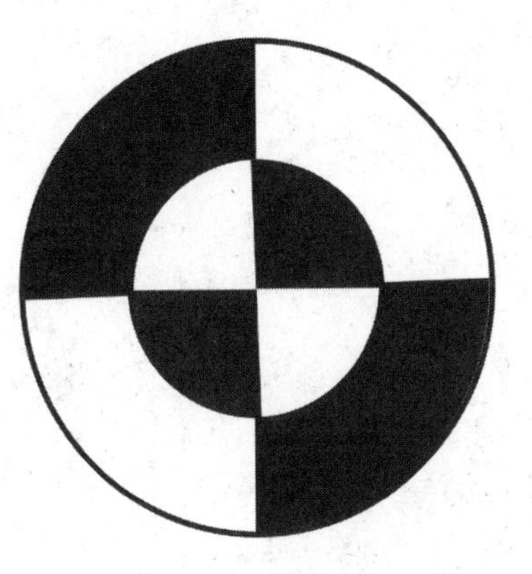

1. Is Riddian's shape about one-half shaded? How do you know?

2. Estimate to shade about one-half of the circle in an unusual way.

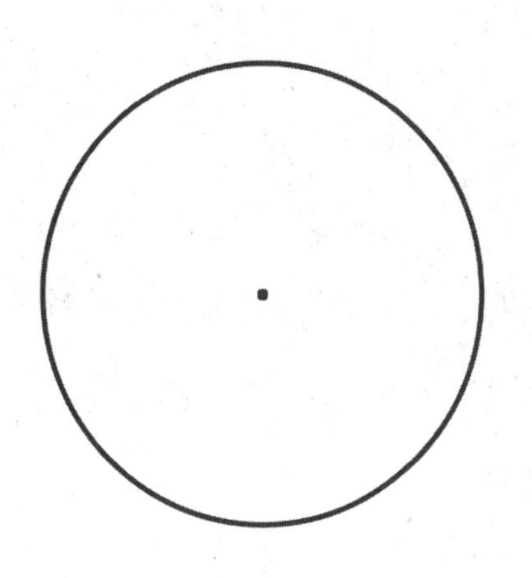

Name ______________________________ Date ____________________

List some games we played today in the chart below. Place a check mark in the box that shows how you felt about your level of fluency as you played each activity. Check off the last column if you would like to practice this activity over the summer.

Activity	I still need some practice with my facts.	I am fluent.	I would like to put this in my summer activity book.
1.			
2.			
3.			
4.			
5.			
6.			
7.			
8.			

Name ______________________________ Date ________________

What fluency activity helped you the most in becoming fluent with your multiplication and division facts this year? Write three or four sentences to explain what made it so useful.

There are 9 bicycles and some tricycles at the repair shop. There are 42 total wheels on all the bicycles and tricycles. How many tricycles are in the shop?

Read Draw Write

Name ______________________________   Date ________________

Complete a math activity each day. To track your progress, color the box after you finish.

Summer Math Review: Weeks 1–5

	Monday	Tuesday	Wednesday	Thursday	Friday
Week 1	Do jumping jacks as you count by twos from 2 to 20 and back.	Play a game from your Summer Practice booklet.	Use your tangram pieces to make a picture of your summer break.	Time how long it takes you to do a specific chore, like making the bed. See if you can do it faster the next day.	Complete a Sprint.
Week 2	Do squats as you count by threes from 3 to 30 and back.	Play a game from your Summer Practice booklet.	Collect data about your family's or friends' favorite type of music. Show it on a bar graph. What did you discover from your graph?	Read a recipe. What fractions does the recipe use?	Complete a Multiply by Pattern Sheet.
Week 3	Hop on one foot as you count by fours from 4 to 40 and back.	Create a multiplication and/or division math game. Then, play the game with a partner.	Measure the widths of different leaves from the same tree to the nearest quarter inch. Then, draw a line plot of your data. Do you notice a pattern?	Read the weight in grams of different food items in your kitchen. Round the weights to the nearest 10 or 100 grams.	Complete a Sprint.
Week 4	Bounce a ball as you count by 5 minutes to 1 hour and then to the half hour and quarter hours.	Find, draw, and/or create different objects to show one-fourth.	Go on a shape scavenger hunt. Find as many quadrilaterals in your neighborhood or house as you can.	Find the sum and difference of 453 mL and 379 mL.	Complete a Multiply by Pattern Sheet.
Week 5	Do arm swings as you count by sixes from 6 to 60 and back.	Draw and label a floor plan of your house.	Measure the perimeter of the room where you sleep in inches. Then, calculate the area.	Use a stopwatch to measure how fast you can run 50 meters. Do it 3 times. What was your fastest time?	Complete a Sprint.

Name ______________________________ Date ______________

Complete a math activity each day. To track your progress, color the box after you finish.

Summer Math Review: Weeks 6–10

	Monday	Tuesday	Wednesday	Thursday	Friday
Week 6	Alternate counting with a friend or family member by sevens from 7 to 70 and back.	Play a game from your Summer Practice booklet.	Write a story problem for 7 × 6.	Solve 15 × 4. Draw a model to show your thinking.	Complete a Multiply by Pattern Sheet.
Week 7	Jump forward and back as you count by eights from 8 to 80 and back.	Play a game from your Summer Practice booklet.	Use string to measure the perimeter of circular items in your house to the nearest quarter inch.	Build a 4 by 6 array with objects from your house. Write 2 multiplication and 2 division sentences for your array.	Complete a Sprint.
Week 8	Do arm crosses as you count by nines from 9 to 90 and back. Teach someone the nines finger trick.	Create a multiplication and/or division math game. Then, play the game with a partner.	Write a story problem for 72 ÷ 8.	Measure or find the capacity in milliliters of different liquids in your kitchen. Round each to the nearest 10 or 100 milliliters.	Complete a Multiply by Pattern Sheet.
Week 9	Jump rope as you count up by tens from 280 to 370 and back down.	Find, draw, and/or create different objects to show one-third.	Go on a shape scavenger hunt. Find as many triangles and hexagons in your neighborhood as you can.	Measure the weight of different produce at the grocery store. What unit did you measure in? What are the lightest and heaviest objects you weighed?	Complete a Sprint.
Week 10	Count by sixes starting at 48. Count as high as you can in one minute.	Draw and label a floor plan of your dream tree house.	Find the perimeter of a different room in your house. How much smaller or larger is it compared to the perimeter of the room where you sleep?	Show someone your strategy to solve 8 × 16.	Complete a Multiply by Pattern Sheet.

Credits

Great Minds® has made every effort to obtain permission for the reprinting of all copyrighted material. If any owner of copyrighted material is not acknowledged herein, please contact Great Minds for proper acknowledgment in all future editions and reprints of this module.